DINOSAURS

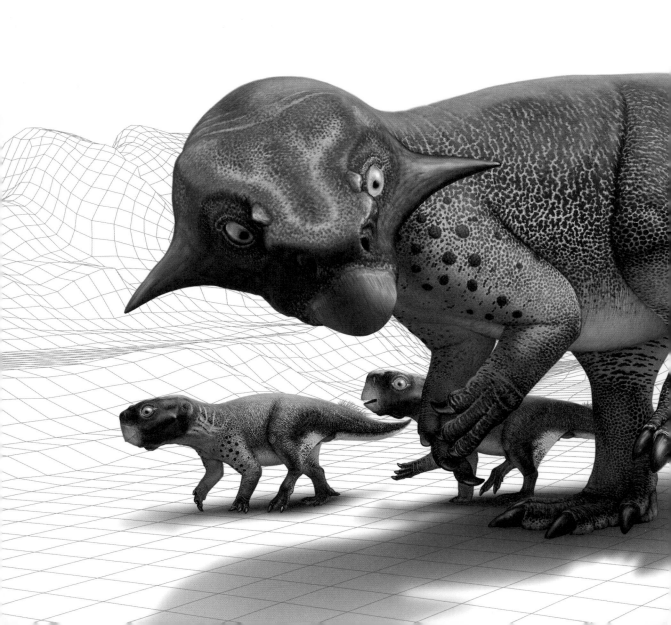

MICHAEL J. BENTON

DINOSAURS

NEW VISIONS OF A LOST WORLD

With illustrations by **Bob Nicholls**

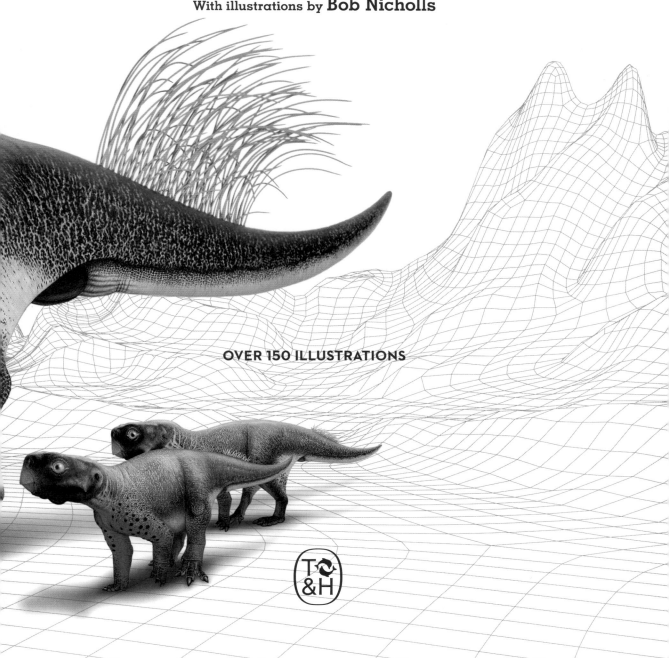

OVER 150 ILLUSTRATIONS

T&H

CONTENTS

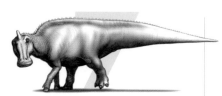

114 EDMONTOSAURUS

A Late Cretaceous hadrosaurid dinosaur that was especially common in North America.

126 EOMAIA

A small Early Cretaceous mammal with fur.

140 SALTASAURUS

This Late Cretaceous sauropod was the first to show fossilized evidence of armour-plated skin.

154 PSITTACOSAURUS

This Early Cretaceous ceratopsian dinosaur and its nests are so abundant in the fossil record that we can visualize it from infancy through to adulthood.

168 KULINDADROMEUS

A Middle to Late Jurassic dinosaur whose skin was covered with both protofeathers and scales, giving insights into the evolution of feathers.

178 STENOPTERYGIUS

An Early Jurassic ichthyosaur that was cleverly countershaded to camouflage itself from its prey.

192 BOREALOPELTA

An Early Cretaceous ankylosaur with red armour plates.

206 ANUROGNATHUS

A Middle to Late Jurassic pterosaur with an unusually short tail that allowed greater manoeuvrability when hunting.

220 TUPANDACTYLUS

An Early Cretaceous pterodactlyoid pterosaur with a distinctive and colourful head crest.

INTRODUCTION

A Late Jurassic pterosaur, *Anurognathus.* **This was a highly manoeuvrable, insect-eating pterosaur (flying reptile).**

Scary reptiles or giant fluffballs?

The monster bears down on the fleeing jeep, the earth shaking as its feet come down with thunderous steps; it snatches up an outhouse in its jaws, grabs the nervous lawyer and swallows him down in one gulp. So we meet *Tyrannosaurus rex* in Steven Spielberg's *Jurassic Park* (1993). It is a truly terrifying sight: all shining, scaly skin, talons and teeth. This image of the dinosaur is so widespread we often don't ask how we know they looked like that. In fact, we now know that they didn't look much like that at all.

Of course, we got some things right. Flesh can be added to the bones with some confidence because muscles leave marks on the bone and, in any case, all tetrapods (four-limbed animals) have pretty much the same muscles; long before anyone had thought about dinosaurs, anatomists noticed that humans and horses, chickens and frogs all have the same muscles in their arms, legs and jaws. They may develop differently depending on whether the animal is a fast runner, or a flyer, or has weak or strong jaws, but their framework of muscles and bones are fundamentally the same. Certain behaviours, like running and feeding, can be determined from simple, common-sense observations. For example, carnivores have sharp teeth like steak knives, while herbivores have blunter, sometimes peg-like teeth. Some dinosaurs moved on all fours, and their arms and legs are of similar lengths, whereas others were bipeds, and so have much longer hind legs. Some, like *T. rex*, have such silly little arms that they were surely never used in locomotion.

But what about the skin? The *Jurassic Park* dinosaurs have reptilian, scaly skin. In 1993, this was all right; now it is not. We don't know much about the skin of *T. rex*, but we do know that tyrannosauroids had feathers. First, the small tyrannosauroid *Dilong* from China was described in 2004 with feathers. At that time, it wasn't clear whether all tyrannosaurs had feathers, or just the small ones – the monsters like *T. rex*, it was argued, could have been entirely scaly. But then another Chinese tyrannosaur, *Yutyrannus*, was described in 2012 with feathers, and this was a giant, some 9 metres (30 feet) long, comfortably in the size range of *T. rex*, at 9 to 12 metres (30 to 40 feet) in length.

In spite of this evidence, by the time the *Jurassic World* sequels premiered in 2015 and 2020, the film makers still preferred to show a scaly *T. rex*. This is a pity, because the *Jurassic Park* films were supposed to be scientifically accurate. However, as one of the producers said, 'We want the dinosaurs to look scary, and that means big teeth and scaly skin. A feathery *T. rex* would just look like an overgrown chicken.' So in a sense, the answer is that dinosaurs have looked how we want them to look, and glancing back through a collection of dinosaur books shows us how they have changed their appearance over the years. Is it all just fashion, though?

Picturing the dinosaur: lizard-like or mammal-like?

The first dinosaurs to be discovered and named were found in England. These were the Jurassic flesh-eater *Megalosaurus*, named in 1824, the Cretaceous plant-eater *Iguanodon*, named in 1825, and the armoured Cretaceous *Hylaeosaurus*, named in 1833. In the 1820s and 1830s, scientists struggled to understand what these huge animals might have been. Some pictured them as giant lizards: one reconstruction of *Iguanodon* had it measuring some 60 metres (200 feet) in length, far larger than its true length of 6 to 8 metres (20 to 26 feet), and with a disproportionately long tail. Others thought these beasts were

mysterious – and massive – kinds of crocodiles, but the fact many of them, like *Iguanodon* and *Hylaeosaurus*, were herbivores caused some problems for this hypothesis. Finally, in 1842, the brilliant, but controversial, biologist Richard Owen (1804–1892) realized these giant bones did not come from either overgrown lizards or crocodiles, but from something else altogether: an unknown taxonomic group. Noting that they had four or more vertebrae in their hip region, unlike the two seen in modern-day reptiles, and because of their huge size he named this family the *Dinosauria*: 'fearfully great reptiles'.

In many ways, Owen was a typical figure of the Victorian establishment; in his photographs he looks austere and grim. He was a friend of Prince Albert, sharing his interest in extending the role of science in public affairs and in the

An early vision of the dinosaur, one of the first attempts ever to provide an image of these ancient beasts. 'Lizard', the frontispiece to George Fleming Richardson's *Sketches in Prose and Verse, containing visits to the Mantellian Museum, descriptive of that collection,* published in 1838.

growth of the British Empire, and was appointed as science tutor for Queen Victoria's children. In 1851, on Prince Albert's initiative, a Great Exhibition was held to showcase the technological and scientific strengths of the economy of Britain and the Empire. It was housed in an enormous glass house built with a steel frame that covered much of Hyde Park. After the Exhibition had finished, the massive greenhouse was dismantled and rebuilt in a south London suburb that was renamed Crystal Palace in honour of the structure. Owen was invited to help create the setting for the new visitor attraction, and he conceived the idea of constructing ancient landscapes in the gardens. He wanted to show the people all the new discoveries by British geologists and palaeontologists – and of course the mineral wealth of the British Isles, especially the coal and

Following pages: Benjamin Waterhouse Hawkins's imposing life-sized model of the theropod (flesh-eating) dinosaur *Megalosaurus*, constructed in 1853 from brick, steel and concrete, and showing Richard Owen's vision of dinosaurs as warm-blooded, rhinoceros-like reptiles.

iron that drove the Industrial Revolution, but in the context of what this showed about the ancient Earth. He reconstructed a Carboniferous swamp, the origin of Britain's rich coal deposits, populated by great amphibians and dragonflies as large as seagulls.

In reconstructing the dinosaurs, Owen reasoned, rightly as it happens, that they had been warm-blooded. This was a startling idea in the 1850s, when dinosaurs were seen as giant reptiles, and thus cold-blooded and slow-moving, like huge, torpid crocodiles. But Owen saw their success as the 'ruling reptiles' of the Mesozoic, and therefore, he reasoned, they must have been warm-blooded, and so he envisaged them like mammals. His *Iguanodon* resembles a rhinoceros, with a massive body supported on four limbs, a huge head and a horned nose.

View of the relocated Crystal Palace exhibition with Richard Owen's fantastical dinosaur reconstructions in the foreground, by the London printer George Baxter.

Megalosaurus was also shown as a hippo-sized predator, and *Hylaeosaurus* as an armoured behemoth. Had he waited for more complete skeletons, Owen might have got it right. He was an accomplished comparative anatomist, meaning he understood the detail of the skeletons and muscles of many living animals, and could relate these to extinct forms. Indeed, for many years, he received deceased animals from London Zoo, and cut them up and described their anatomy in detail. Owen's vision of the dinosaurs was revolutionary, and these models remain a testament to the extent of knowledge at the time.

Complete dinosaur skeletons were found in the United States from 1855 onwards, and these showed that in fact many of the known species were bipedal, including *Iguanodon* and *Megalosaurus*. One of these more or less complete

skeletons was *Hadrosaurus*, from the Cretaceous of New Jersey. When Joseph Leidy (1823–1891) named *Hadrosaurus* in 1858, he realized this 7-metre (23-foot) long animal must have been a biped: its hind legs were more than twice the length of its arms, and so it must have hoisted its body upright, and run along on just its hind legs. Perhaps, when feeding low on the ground, it could crouch down and use its arms in walking too. When the skeleton was put on show at the Philadelphia Academy of Natural Sciences in 1868, it was the first ever museum mount of a dinosaur, and it was upright. Quite unlike the scaly rhinoceroses of Owen's imagination!

This set the scene for a rush of new dinosaur discoveries from North America, from 1865 to 1900, a time sometimes called the 'great dinosaur rush' or the 'bone wars'. Two palaeontologists in particular competed to be the first to find and name the best examples. Edward Drinker Cope (1840–1897) was associated with the Philadelphia institution, and indeed was much influenced by Leidy and his *Hadrosaurus*. He learnt his palaeontology through study, touring the great museums of Europe, and used his family wealth to finance expeditions to the American West, where the railroad crews were unearthing the great bones of mastodons and other giant mammals, as well as Jurassic and Cretaceous dinosaurs. His rival was Othniel Charles Marsh (1831–1899), a professor at Yale University, where his wealthy uncle, George Peabody, founded the Peabody Museum in 1866. Cope and Marsh had the finances and the ambition, and their field crews packed great boxes of bones to be sent back by rail to the East Coast. They opened the boxes and their technicians cleaned up the bones, and Cope and Marsh would write descriptions of the new beasts, providing them with names, at great speed. Thanks to them, we have *Allosaurus*, *Brontosaurus*, *Stegosaurus*, *Triceratops* and dozens more classic American dinosaurs.

At this time, the most influential dinosaur artist was Charles Knight (1874–1953), who provided museums and publishers with his services. He began painting dinosaur scenes in the 1890s, in Cope and Marsh's later years, and used their great discoveries as a basis for his art, as well as taking inspiration from the dinosaur galleries of the American Museum of Natural History in New York. He also spent a great deal of time studying animals in the zoo and noting how their limbs and muscles moved. He applied this knowledge to his dinosaur paintings, and they felt alive. He was soon employed to paint colour murals of dinosaurs, and other extinct animals, as if in life, and his paintings were hugely influential, reproduced worldwide in books and in popular magazines such as *National Geographic*.

Opposite:
A skeletal restoration of *Hadrosaurus foulkii* in the United States National Museum (now the Arts and Industries Building), 1880, based on the original in the Philadelphia Academy of Natural Sciences.

Following pages:
Leaping Laelaps, Charles Knight, 1897. Knight's careful study of the anatomy and behaviour of living animals enlivened his sketches of extinct species.

CHAS. R. KNIGHT
97

Revolution in dinosaurian posture

One image – a simple pencil drawing, made in 1969 – can be credited with driving a revolution in our perception of the dinosaurs. Up till then, dinosaurs had been seen as awkward, bulky and slow-moving, but after this drawing, artists began to show them as balanced and sleek – and fast.

The drawing shows *Deinonychus*, a medium-sized predatory dinosaur known from several complete skeletons excavated in the 1960s in the Early Cretaceous of Montana. John Ostrom (1928–2005) of Yale University had led the excavations, and in 1969 he published a beautiful and detailed anatomical study of the animal. Ostrom realized that *Deinonychus* was a most unusual dinosaur – it was slender and built for speed. It had powerful, long arms, quite unlike the reduced forelimbs of *Tyrannosaurus* and many other large flesh-eaters, and the feet had a most unexpected switch-blade adaptation: the claw on the second toe was huge and sickle-like. When the animal was walking or running, Ostrom reasoned, it must have held the claw up, clear of the ground, otherwise it would have worn down. He showed how the toe bones allowed the claw to be pulled right back, and then to sweep downwards through an arc of 180 degrees.

Ostrom postulated that *Deinonychus* balanced on one foot, perhaps using its long arms to steady its position, and deployed the slashing claw to tear the flank of its prey. Indeed, he had noted that its most likely prey, the herbivore *Tenontosaurus*, was ten times larger than *Deinonychus*, and so the wiry predator must have evolved special tactics to weaken and kill its prey. So, rather as wolves harass and nip at the hamstrings of huge caribou in the Canadian winter, perhaps *Deinonychus*, whether hunting singly or in packs, could bring down a huge herbivore by guile rather than brute force.

This kind of balance and intelligence was not expected in a dinosaur. What's more, Ostrom realized that *Deinonychus* was a close relative of birds. In every aspect of its skeleton, it resembled *Archaeopteryx* – slender body, powerful hindlimbs, long arms and powerful hands, long, thin bony tail, and pointed snout with small, recurved teeth lining the jaws. This is just what Victorian anatomists such as Thomas Huxley (1825–1895) had claimed exactly one century earlier.

Huxley was a great supporter of Charles Darwin's theory of evolution, and he argued that *Archaeopteryx*, which had been found in 1861, was simply a small dinosaur in bird's clothing. After that, palaeontologists had lost their nerve somewhat, and couldn't see that a feathered bird such as *Archaeopteryx* had evolved directly from a theropod (flesh-eating) dinosaur. Birds, they said, must have originated from much more ancient, if rather obscure, ancestors over 50 million years older than *Archaeopteryx*, in the Triassic.

Ostrom put everyone right. *Deinonychus* was clearly hugely bird-like, and even though it used its long, sinewy arms for grappling with prey rather than flying, Ostrom probably dreamt that it might have been covered with feathers. This he did not say, because he was a careful scientist, but he lived to see his ideas vindicated by the first reports of the feathered dinosaur *Sinosauropteryx* in 1996.

But the 1969 drawing of *Deinonychus* was not by Ostrom, but by his student Bob Bakker (b. 1945). Bakker was, and is, an iconoclast, and as a student, he could hardly contain his artistic and interpretive skills. The pencil sketch has the advantages of simplicity and clarity of execution. Here is a dinosaur in a hurry. Abandoning the kangaroo-style posture of *Hadrosaurus* and other bipeds, the skeleton is balanced. Of course, that has to be true. As a biped, you either stand straight upright as humans do, or you balance your body fore and aft over the limbs. Birds have short tails, and so the weight of the body is distributed by having about half the trunk in front and half behind the knees. Dinosaurs, with their long tails, similarly had to balance like a see-saw, with head and torso out front and hips and tail behind. While running, as here, the backbone was held more or less horizontally.

Bob Bakker's revolutionary 1969 sketch of *Deinonychus*, based on a skeleton found by John Ostrom, showing its sickle like claw and superior balance – a creature built for speed.

Bakker '69

Working out the stats:
top speeds and bite forces

Correcting the dinosaurs' posture was common sense. The need for balance during locomotion is a simple observation in physics: if you don't balance, you fall over. The mechanics at play today worked just the same in the Jurassic.

Palaeontologists have built on this kind of common-sense approach to expand the scope of what can be tested about the lives of extinct organisms. For example, the speed at which an animal runs is determined by two things – muscle volume and stride length. Muscle volume is directly proportional to the maximum speed that can be achieved: fast runners, such as Olympic sprinters, have massive leg muscles. The same is true of stride length: as you transition from a slow walk to a jog, and then to an all-out sprint, your stride length doubles and doubles again. In fact, both of these physical relationships provide simple formulae that can then be applied to any living or extinct animal.

This method was first proposed by Robert McNeill Alexander (1934–2016), professor of biomechanics at the University of Leeds, who set out to test his footprint–speed formula by putting a series of animals – and his family – through their paces on a North Norfolk beach. He found that his calculation worked for dogs, racehorses, humans, ostriches… indeed, everything. Providing you know the animal's stride length and the length of its leg, you can accurately predict the speed of locomotion. Alexander reasoned the same must also apply to extinct animals, and published a series of speed estimates for dinosaurs.

But could we trust these estimates? Theoretically, the formula – and its application to extinct creatures – is sound. But additional confirmation came from studies of the muscle volume–speed formula by John Hutchinson, professor of animal locomotion at the Royal Veterinary College near London. His independent calculations of leg muscle volumes and inferred speeds for dinosaurs gave the same figures as McNeill Alexander's track estimates: *T. rex* could run at 27 kilometres (17 miles) per hour, a comfortable speed for driving round a built-up area, but in no way equivalent to a racehorse, which can achieve speeds of 70 kilometres (43 miles) per hour. Using modern animals as analogues for extinct animals is a method that has also been applied to dinosaur feeding – after all, who could resist finding out how hard a *T. rex* bites?

Bite force is the force an animal can exert on a food item using its jaws. The jaw has a hinge at the back, and the force is provided by the muscles that run from the jaws to the sides of the skull. Bigger muscles mean bigger force. The Great White shark can exert a bite force of 18,000 Newtons – quite impressive, when you consider that the best a human can achieve is about 1,200 Newtons. There are about 10,000 Newtons to 1 tonne, so the Great White bites the unfortunate swimmer with a force equivalent to a weight of 1.8 tonnes bearing down on them.

Still, *Tyrannosaurus rex* puts these living animals to shame. Its bite force has been calculated as 35,000 to 57,000 Newtons, or 3.5 to 5.7 tonnes. These figures can be calculated in two ways – one quite hi-tech, the other more like a kitchen-sink experiment. The hi-tech calculation relies on estimates of the material properties of the bone of the *T. rex* skull and estimates of muscle volumes. The kitchen-sink approach is based on a bit of good luck – finding material evidence of the bites. One such find, a fossil *Triceratops* hip bone bitten by a *T. rex*, shows a tooth impression 3 cm (1¼ in.) deep. Experiments with modern cow bone and a steel tooth show the force needed to drive the tooth 3 cm deep into the bone was a mere 6,400 Newtons, much less than the maximum force *T. rex* could exert. Similar calculations have been done for the giant shark Megalodon, denizen of the seas 5 million years ago (long after the dinosaurs), but 16 metres (52½ feet) in length (the Great White is 5 metres, or 16½ feet) and with a bite force four times greater – as much as 5 tonnes, similar to T. rex.

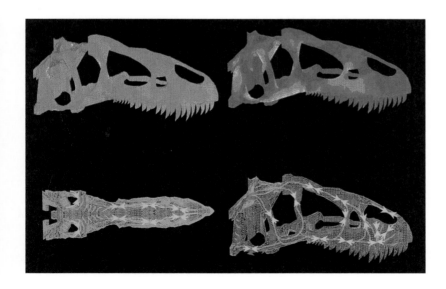

A computer modelling of the skull of *Allosaurus fragilis*, a predatory dinosaur from the Late Jurassic period. This model was created to study the mechanical properties of the skull by Emily Rayfield. The lower images show forces of compression (yellow arrows) and tension (red arrows), as seen from the side (lower right) and below (lower left). The model is based on a computed tomography (CT) scan of an *Allosaurus* skull and the results suggest that *Allosaurus* had a fast, downward slashing bite rather than the powerful, crushing action found in its relative *Tyrannosaurus rex*.

Applying science

These calculations about dinosaur speed and biting are *scientific*, meaning they are built on two assumptions common to all scientific endeavour. One of these principles is called uniformitarianism: we assume that the laws of physics are the same everywhere and at all times. McNeill Alexander assumed that extinct animals obeyed the footprint–speed formula, and we assume that dinosaur jaws worked just like those of modern animals. The second principle is corroboration: in calculating the speed and bite force of *T. rex*, palaeontologists corroborated their findings by following two independent approaches.

Such advances are happening in every area of palaeontology at a remarkable rate. Every year, hundreds of papers about dinosaurs are published in scientific journals. If you simply read the press reports, you would think that most of these are announcing new species. The headlines are common enough: 'New dinosaur species from China is largest/ smallest/ widest/ shortest ever'. Palaeontologists do this. Our job is to document the life of the past. But palaeobiology is much more fun, and much smarter. In fact, half the papers published about dinosaurs investigate how dinosaurs lived their lives, and these papers often use lateral thinking – some smart and unexpected insight – to establish something about these largest-of-all land animals.

Scientists are drawn to extremes. For centuries, biologists have studied every aspect of the lifestyle of elephants, to understand how these huge animals function. Their great size presents unique challenges: they have to establish vast territories in order to be able to find enough food to power their enormous bodies; their skeletons have evolved to support their huge weight; and their physiology functions to enable them to be active, digest enough food, and not overheat in the hot climates of Africa or India. For a long time, people of all nations wondered at these huge animals, and scientists declared they were at the limits of possibility; nothing could be larger and still survive.

And yet dinosaurs were larger – many of them ten times larger (weighing 50 tonnes to the 5 tonnes of an elephant) – and they survived in a world just like ours in most regards. We can't solve the question of their adaptations by saying gravity was less in the Jurassic than today, or that they slouched about under water to keep their bodies afloat. There is no evidence for that. They truly pushed the limits beyond anything we see today. How did they do it?

The true giant dinosaurs, the sauropods (large herbivores) like *Diplodocus* and *Brachiosaurus*, evolved a smart combination of reptilian and mammalian-like characteristics. A 5-tonne elephant eats 225 kilograms (500 pounds) of food each day, and spends more than 12 hours grabbing and chomping to achieve this; nine-tenths of this food intake is used by the elephant to maintain a constant body temperature. A 50-tonne sauropod required the same amount of food, because it did not use internal physiological mechanisms to control its body temperature (see pp. 152–53). Second, sauropods laid quite small eggs and probably abandoned them – one mother could produce twenty or more eggs each year, and if only one or two survived to adulthood, that would be enough to ensure the survival of the species. Mother elephants carry their single baby for 18–22 months, and giving birth and caring for the calf are risky undertakings. The giant dinosaurs could be ten times the size of an elephant because they put in only one-tenth of the effort in feeding and childcare.

The final frontiers

So where are dinosaur studies now? We can learn about dinosaur feeding and locomotion. We can say quite a lot about dinosaur parenting (or lack of it) – dinosaur nests and eggs are found regularly, in some cases with tiny embryos inside the eggs. We can learn about dinosaurian ecology by examining sites of exceptional preservation that reveal the plants, insects and other animals that lived side by side with dinosaurs, allowing palaeoecologists to reconstruct food webs, drawing complex diagrams of energy flow through an ancient ecosystem.

The focus of this book is the appearance of dinosaurs – their colours, patterns, and the nature of their skin coverings. At one time, there was no call for such a book, because all dinosaurs were thought to have had scales or osteoderms (bony plates set into the skin), just like modern reptiles. Scales, seen in fishes and reptiles, are typically made from the transparent protein keratin; osteoderms appear on many crocodiles. Of course many dinosaurs had osteoderms too, and we know this because, being made from bone, they are commonly preserved. Thus, in the works of early palaeoartists such as Charles Knight, dinosaurs were all scaly or covered in bony osteoderms, and mostly in dull colours of greens, browns and khakis.

Yet this was never so.

John Ostrom and Bob Bakker hinted at a new world of dinosaurs in 1969. Some theropods, they hypothesized, around the time of the origin of birds, were active, warm-blooded, and might even have had feathers.

Microraptor, a four-winged dinosaur, had long pennaceous feathers on its arms and legs, a type common to most modern birds. These made *Microraptor* aerodynamic and would have helped in gliding and perhaps even powered flight.

Feathers: fashionable or functional?

Ostrom and Bakker's predictions were vindicated from 1995 onwards, when an astonishing assemblage of fossils was discovered in the north of China. Each was more exquisitely preserved than the last, and most showed that small theropod dinosaurs were equipped with a remarkable array of feather types. At first, these fossils were controversial – surely, many asked, dinosaurs could not have feathers, because feathers are a bird thing.

But indeed they did, and all kinds of feathers. Soon, the Chinese fossils showed every type of feather seen in modern birds, plus a few more. And it wasn't just the smaller theropods; some large ones, such as the 9-metre (30-foot) long tyrannosaur *Yutyrannus* had feathers. Further, it seems feathers of one sort or another were present not only in theropods and birds, but also in ornithischians, plant-eating dinosaurs that are not at all closely related to theropods and birds. This suggested strongly that all dinosaurs had feathers, and that feathers had originated long before birds.

Feathers brought new insights, because they carry pigments, and in 2010 two teams independently determined their colours and patterns in two species of dinosaurs. Suddenly dinosaurs were no longer all scaly or drab. Some at least had fancy feathers covered with stripes, speckles and spangles. This opened up questions about feather function – were they for warmth, social or sexual signalling, or flight? Obviously flight came third, among birds and their close relatives, but which of the other two came first? The broadest sampling suggests that the first feathers were a dull brownish colour, indicating that insulation was their primary purpose; it is among more advanced forms that we find the long tail feathers, coloured stripes and markings and other structures used for signalling, especially in pre-mating display, in many modern birds.

Colour can also function for camouflage, and the brownish colours of some dinosaurs likely aided them in disappearing among the foliage when a predator had its eye on them. More than that, some dinosaurs showed countershading (that is, a darker back and lighter belly), enabling them to break up their outlines and become seemingly two-dimensional in different conditions of lighting, whether out in the open or among the dappled light streams of a forest.

Here we present, for the first time, a vision of the dinosaur world as near to accurate as can be managed. In the exquisite palaeoart of Bob Nicholls, we see all the inferences about dinosaur posture, behaviour and external appearance drawn together, bringing these long-extinct species to life on the page. These artworks are based on a hard, critical look at the latest fossil evidence and the cutting-edge research taking place in academic institutions all over the world. This is just a beginning – the days of fanciful dinosaur books are over.

Different types of feathers had different purposes. *Anchiornis* could not fly, but its feathers may have helped to regulate its temperature, or to signal its age or sex to other members of its species.

PUTTING THEM IN THEIR PLACES

Geologists work with a standardized time scale, running from the origin of the Earth, 4,567 million years ago, through the time divisions shown here, when fossils were abundant. Dinosaurs, for example, existed in the Mesozoic, originating in the Triassic, flourishing in the Jurassic and Cretaceous, and then dying out at the very end of the Cretaceous, 66 million years ago. Here we show the ages of the fifteen prehistoric reptiles, birds and mammals we include in the book.

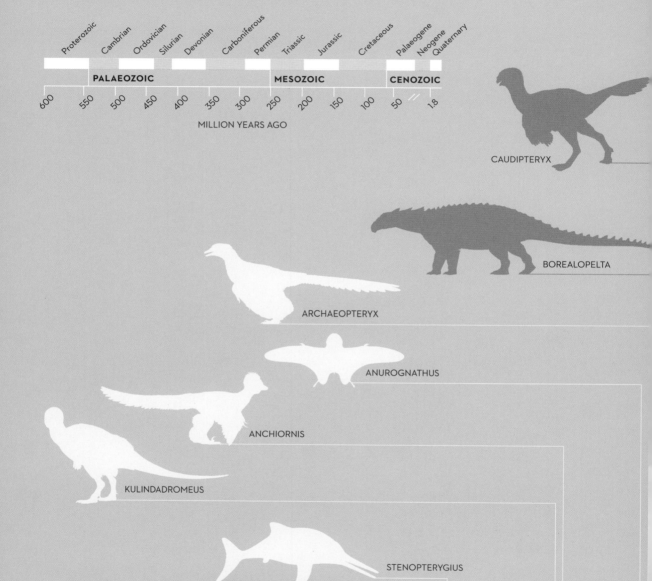

Proterozoic · Cambrian · Ordovician · Silurian · Devonian · Carboniferous · Permian · Triassic · Jurassic · Cretaceous · Palaeogene · Neogene · Quaternary

PALAEOZOIC   **MESOZOIC**   **CENOZOIC**

600 · 550 · 500 · 450 · 400 · 350 · 300 · 250 · 200 · 150 · 100 · 50 · 1.8

MILLION YEARS AGO

CAUDIPTERYX

BOREALOPELTA

ARCHAEOPTERYX

ANUROGNATHUS

ANCHIORNIS

KULINDADROMEUS

STENOPTERYGIUS

TRIASSIC 252–201 MYA   **JURASSIC** 201–145 MYA

MESOZOIC

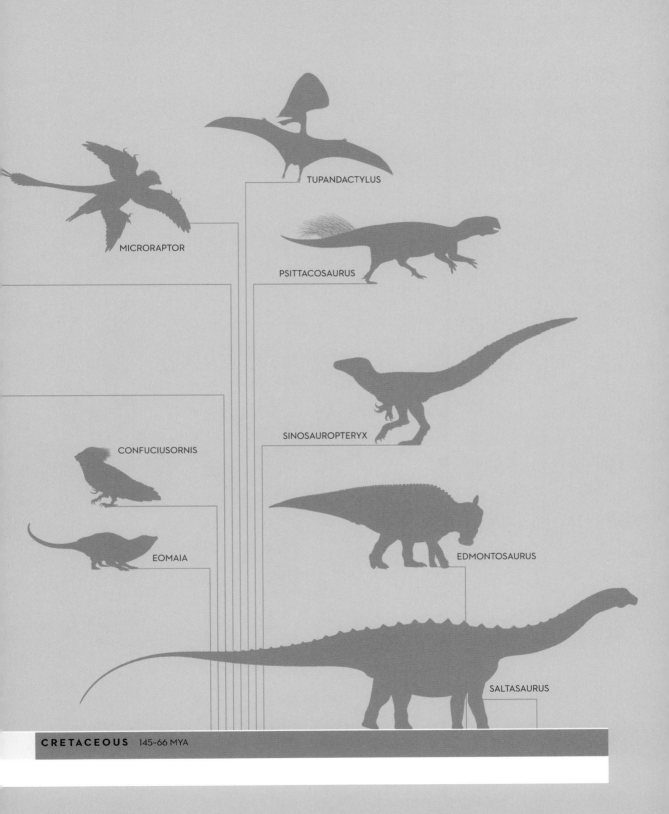

MICRORAPTOR

TUPANDACTYLUS

PSITTACOSAURUS

SINOSAUROPTERYX

CONFUCIUSORNIS

EDMONTOSAURUS

EOMAIA

SALTASAURUS

CRETACEOUS 145–66 MYA

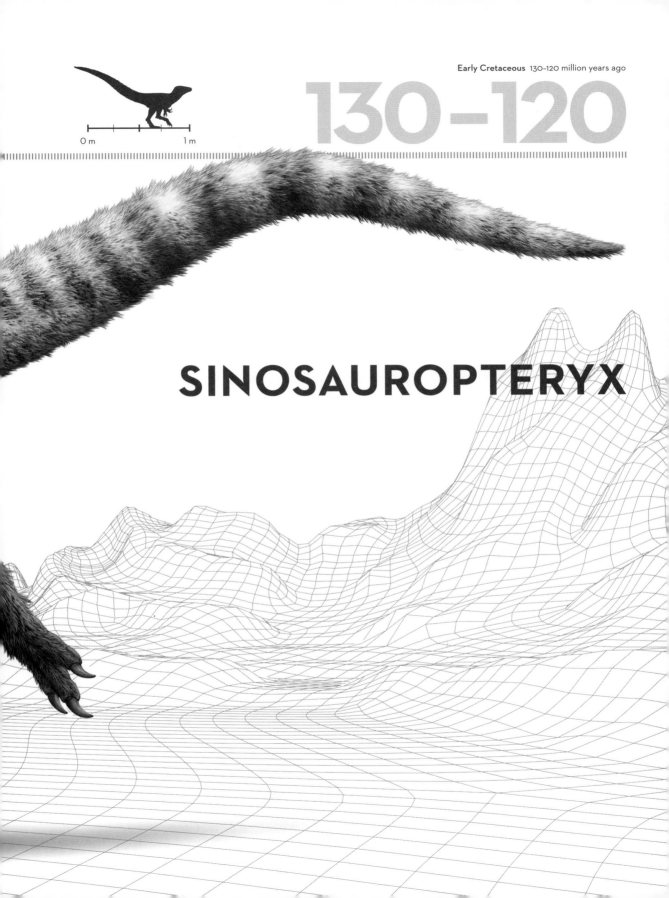

1

0 m   1 m

SINOSAUROPTERYX

Ginger stripes and a bandit mask

Let's go back 125 million years, to the Early Cretaceous, and the open, bushy landscapes of what is now northern China. These woods are alive with birds, mammals, lizards, salamanders, and many other modern-looking animals. Insects buzz among the reeds and bushes that surround sluggish lakes, and volcanoes in the near distance rise above the trees.

You hear rustling, and a fast-moving little dinosaur appears, zig-zagging in pursuit of a lizard. He snaps and the lizard is lunch. The dinosaur is a metre long, covered in a pelt of short, hair-like feathers, and marked by striking patterns. The tail is striped, ginger and white, from end to end, like a fluffy old-fashioned barber's pole. It has a stripy face to match, with a ginger bandit mask highlighting its eyes. This is *Sinosauropteryx*, the first dinosaur found to have feathers – a discovery that sparked a revolution in dinosaur science, beginning in 1996.

Sinosauropteryx is a theropod, the dinosaurian group that includes all the flesh-eating forms plus the birds. As he runs into the open, he freezes in the bright sunlight, and seems to disappear into the landscape of layered brown colours. How can we know the colours and patterns of a dinosaur in such detail? And what were all the colours for?

Clever colour adaptations

We know that colours in modern animals have many functions, and a stripy tail can mean many things. It might be for camouflage, like the tail of a tiger, or for display, like the tail of a lemur.

When we first discovered the *Sinosauropteryx* stripes and their colours in 2010, we ruled out camouflage, because the stripes should then have extended all over the body so that they acted as disruptive coloration, breaking up the outline of the little dinosaur in the dappled sunlight coming through the trees. But *Sinosauropteryx* has a ginger body, with a pale belly. The idea that this was a display structure was more convincing, and a realization that changed our view of dinosaurs yet more, away from reptiles and towards birds. Even though *Sinosauropteryx* sits a long way from birds in the evolutionary tree, these little orange dinosaurs likely hopped around waving their stripy tails to show off and attract mates.

Above:
The *Sinosauropteryx* tail was ginger and white, in stripes of equal widths, and it might have functioned as a warning or sexual display signal, just as in the ring-tailed lemurs of Madagascar today (opposite).

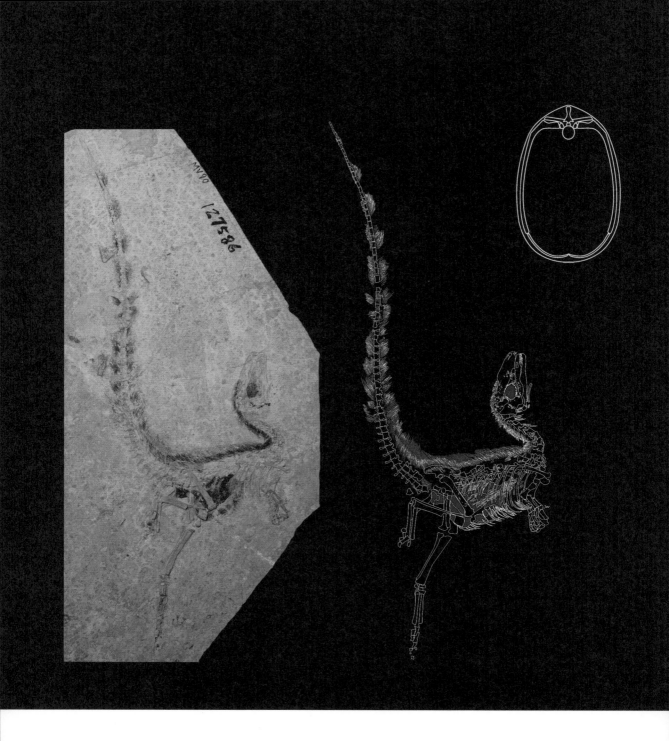

Two fossil specimens of *Sinosauropteryx*, NIGP 127586 (this page) and NIGP 127587 (opposite), in each case showing a photograph (left) and an interpretive drawing (right), highlighting the feathers and preserved internal organs. These dinosaurs are 1.3 m (4¹/₄ ft) and 1.6 m (5¹/₄ ft) long respectively.

The white silhouettes at the top of each page are cross-sections through the rib cages of these two dinosaurs, showing a thinner (left) and fatter (right) individual; their girth matters when reconstructing how far down the side the dark-light colour transition extends in understanding how their countershading

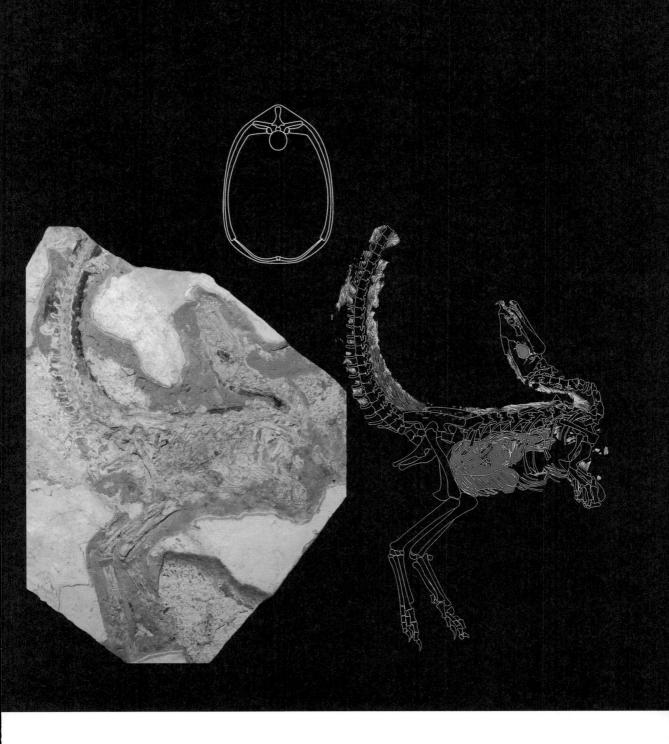

worked in life. Pigment traces around the face region have been interpreted by Fiann Smithwick as evidence for a 'bandit mask', with a slash of dark shading over the eye set in a generally light-coloured face. Traces of coloured feathers along the sides indicate the countershading went quite low. Unusually, both specimens preserve the internal organs quite extensively; inside the rib cages are remains of the stomach and other structures, but they have been reduced to a hard-to-interpret organic mass. The larger dinosaur has remains of a lizard in its stomach, evidence of its last lunch before it died.

Detail of the hands and right foot of *Sinosauropteryx* on the run. Here the focus is the claws, and it is worth recalling that what we see of a claw is the outer sheath and not just the internal bone. Just as humans have fingernails made from the protein keratin, dinosaurs and birds have claws also made from keratin. In these cases, the claw sheath wraps around the claw bone and extends its length, sometimes to as much as twice the length. The foot claw sheaths are worn down as the animal runs about, but the hand claws keep their sharp, cruel curve.

But what about the facial markings? In 2017, a PhD student, Fiann Smithwick, and his supervisor, Jakob Vinther, at the University of Bristol, noted that *Sinosauropteryx* had a 'bandit mask', as they called it: a dark, ginger stripe of colour that surrounded the eyes, and ran back to the angle of the jaw. By comparison with modern mammals such as skunks and raccoons, Smithwick and colleagues suggested this might have been a warning device, advertising that *Sinosauropteryx* packed a nasty weapon – a stinky exudate or, more likely, its fierce, sharp claws. This told larger predators to keep clear; if they attacked, little *Sinosauropteryx* would fight back.

Smithwick and colleagues also identified that *Sinosauropteryx* was countershaded, with a lighter belly region and darker upper body coloration. The position of the countershading line tells us something of this dinosaur's natural habitat; whereas the division between light and dark lies low down the side in other dinosaurs such as *Psittacosaurus* (pp. 154–67), indicating that they lived in densely forested areas, in *Sinosauropteryx* the line of demarcation was higher, suggesting it inhabited brighter, open spaces.

The light and dark bands of these colour patterns can clearly be seen on the feathers in some of the spectacular fossils of *Sinosauropteryx*. But how do we know that the darker colour was ginger?

Revealing prehistoric colours

In 2005, I noticed that the Royal Society was offering a fellowship for a young Chinese researcher to come to the UK, the K. C. Wong Award. I wrote to Zhonghe Zhou at the Institute of Vertebrate Paleontology and Paleoanthropology (IVPP) in Beijing to ask whether he could recommend a bright young person. He thought his colleague Fucheng Zhang would make an excellent candidate, and so we applied and secured the funding, and Fucheng came to join our team at the Bristol Palaeobiology Research Group for a year. Packed among his shirts and socks, he brought some specimens he was studying, including fossil birds with feathers. We were captivated. Could we perhaps visit China, we asked, and even begin studying some of the specimens ourselves?

In summer 2007, three of us visited IVPP, and were treated to a two-week field excursion around all the fossil sites in Liaoning and Hebei provinces, followed by a week studying specimens at the institute. Could we borrow a few specimens to study in more detail back in Bristol? Yes, of course; and so we staggered back to the UK with our portable microscopes, a pile of books and papers about the amazing Liaoning fossils, and some small rock pieces with individual feathers and other samples. Our initial microscopic studies were promising, and we asked Fucheng to collect some specific samples from *Sinosauropteryx* and other Chinese feathered dinosaurs and early birds.

Fucheng Zhang returned to Bristol in November 2008, and we put his samples under the Scanning Electron Microscope. Paddy Orr and Stuart Kearns, my collaborators, had had a hunch – they were sure we would be able to identify minute structures within the material of the fossil feathers. They had already determined that the fossilized feathers were not merely impressions or recrystallized structures. They were three-dimensional, if a little squashed, and there was every chance we could see details of their microstructure.

And indeed we did.

On 27 November 2008 we saw prehistoric melanosomes for the first time, and having seen them in one specimen, we saw them in every specimen in every view. Melanosomes (see pp. 52–55) are capsule-like structures of fixed sizes and shapes buried within the keratin protein that makes the feather (or hair, in mammals), and they proved the *Sinosauropteryx* 'protofeathers' really were feathers and not shredded skin. One problem solved.

They also told us the colour of the feathers, and this was another first. Biologists had established the matching of melanosome shapes and colours of the pigment melanin, which meant we could show that *Sinosauropteryx* had ginger and white feathers in zones around the body. Most of the feathers on top of the head and down the midline of the back were ginger, and the tail had alternating stripes of white and ginger.

We checked more *Sinosauropteryx* specimens, as well as feathers from the dinosaur *Sinornithosaurus* and the bird *Confuciusornis* (pp. 98–113), and they showed the same structures preserved in the same way. We also compared our findings with those in feathers of modern birds. These were harder to see, because the modern feathers were complete, and the outer surface of the feather barbs was simply smooth. Stuart Kearns devised a smart way to dry-freeze feathers of a modern zebra finch in liquid nitrogen and then crack them open with a physical blow. The cracked surfaces through the feather structure mirrored the fractures in the fossil feathers, which had split along a line of weakness in the rock that enclosed the fossil. The interior structures were now laid bare, and we could see the solid structures of the melanosomes, as well as rounded pits where melanosomes had popped out of the keratin feather matrix, leaving a mould behind.

Time to go to press. We wrote our paper in early 2009 and submitted it to *Nature* in July 2009. After three rounds of review by four referees, it was finally accepted in November 2009 and published in January 2010. At last, we could announce that, for the very first time, we knew the colour patterns of a dinosaur, and could use these insights to speculate about dinosaurian behaviour.

This revelation was followed a few weeks later by a parallel study on the feather colours of the dinosaur *Anchiornis* (see pp. 42–55). Both papers created a huge stir, with wide press reporting, and indeed the discovery of how to tell the colour of dinosaur feathers was awarded the accolade of being one of the top ten discoveries in all of science in the decade from 2010–2020 by the Smithsonian Institution. It's not often that palaeontology sits up there, side by side by the Nobel-prize-winning discoveries of fundamental chemistry and physics!

These discoveries are a long way from the early adventures in dinosaur art. And it wasn't even the first time that *Sinosauropteryx* had turned the expectations of the palaeontological community on their head.

Opposite:
Sinosauropteryx snatches lizard *Dalinghosaurus* from a small pool and shakes it. Dinosaurs did not chew their food, so he grabs the prey animal, twists it round so it faces head-first and gulps it down. This is why there is a whole lizard skeleton inside the guts of one of the museum specimens of *Sinosauropteryx* (see page 33). This image shows the ginger-and-white striped tail, the countershading along the side, and the bandit mask. Painting by Bob Nicholls.

Nicholls 2017

The first feathered dinosaur

We saw earlier (pp. 18–19) how John Ostrom had identified that the skeleton of *Deinonychus* was hugely birdlike, and that he and his student Bob Bakker had changed perceptions by showing these small dinosaurs were fast and warm-blooded. But it wasn't until *Sinosauropteryx* was presented to the world in 1996 that their suspicions were vindicated.

This event was at a conference, and not at first through a published paper. I was present as an inconspicuous delegate at the 56th Annual Meeting of the Society of Vertebrate Paleontology, held that year at the American Museum of Natural History in New York. The excitement was not in any of the formal talks or posters, but in the bars and coffee rooms. A small delegation had come from China for the first time; earnest young men in dark suits, perhaps a little nervous at their first appearance at the conference, but keen to meet the leading palaeontologists in person. Phil Currie, the famous dinosaur professor at University of Alberta and founder of the great Royal Tyrrell Museum of Palaeontology, made the introductions, and they quickly homed in on Ostrom.

Currie had been excavating in Mongolia and flew to Beijing to study dinosaur specimens in September 1996. Zhiming Dong, then the senior dinosaur palaeontologist at IVPP, told him about an unusual new fossil bird. The specimen had been found in August 1996 near Sihetun in Liaoning Province by farmer Yumin Li, who recognized that he had found something amazing. Showing great business acumen, Li split the specimen into two – the slab and counterslab – and sold these to two separate institutions, the National Geological Museum (NGM) in Beijing and the Nanjing Institute of Geology and Palaeontology (NIGPAS).

Perhaps suspecting that there might be some competition to name the new beast, two of the NGM palaeontologists, Qiang Ji and Shuan Ji, wrote a short article and published it in *Chinese Geology*, the house journal. The article was in Chinese and it was not illustrated, but Ji and Ji gave the new fossil a name, *Sinosauropteryx prima*: 'Chinese first reptile wing'. This established the NGM half of the specimen as the type example. Ji and Ji were convinced they had a fossil bird, because it had feathers.

One of the original specimens of *Sinosauropteryx*, NIGP 127586, highlighting the skeleton. Internal organs can be seen inside the rib cage, and the fringe of feathers over the top of the skull, down the middle of the back, and in tufts along the tail.

On Wednesday 25 September, Currie was invited to see the fossil during a press announcement of the short paper that named *Sinosauropteryx*. In a conference room at the NGM, he was shown specimen after specimen, each in a silk-covered box. Currie recalled the occasion in a letter to me, years later. 'First he shows me a beautiful *Psittacosaurus*, then works his way through insects, fish, lizards, a mammal with fur, birds… By now I am convinced that they are not going to show me the *Sinosauropteryx*… Ji Qiang then opens without announcement another box and my "jaw drops" as I see the specimen for the first time. It is gorgeous as a skeleton, but within milliseconds I shift my view to the "feathers" which I now know are not dendrites or fungal growths. I am flabbergasted as I go over the specimen with only a magnifying glass. The press leaves after about an hour.'

The new specimen was a complete skeleton of a slender dinosaur on a slab measuring about 1 m (3¼ ft) long. The animal was posed balletically, with one leg and foot fully extended, the other tucked up behind, the arms hunched into the chest, the tail straight up in the air and the head and neck bent back. Currie was quoted by the *New York Times*: 'I was bowled over.'

Currie flew home, but, 'Within a few days of getting back to Canada… I had a surprise phone call from another Chinese colleague and friend – Chen Peiji. He told me that the counterpart was in fact in Nanjing (the institute had bought it without knowing that the other half had been sold to Beijing). I was invited to work with Chen on this specimen! We agreed to meet at the SVP meetings in New York the following week and that I would go to China as soon as possible after that. On Oct. 12, I left for New York.'

In New York, Phil Currie and his Chinese colleagues showed the photographs to John Ostrom. This was what Ostrom had always dreamed about – a dinosaur with feathers. He reported later that these photographs left him 'in a state of shock'. In 1998, Chen and colleagues from NIGPAS, together with Currie, described their specimen as well as its matching counterpart, but this time in the international journal *Nature*, and with convincing colour illustrations. Now the world paid attention.

The *Sinosauropteryx* fossil site, near Sihetun in Liaoning Province, northeast China. This lakebed geological formation is 30 m (100 ft) thick and covers 50 square kilometres (20 square miles). Fossil collectors dig through the thin limestone beds, and even excavate tunnels in the search for special fossils.

2

ANCHIORNIS

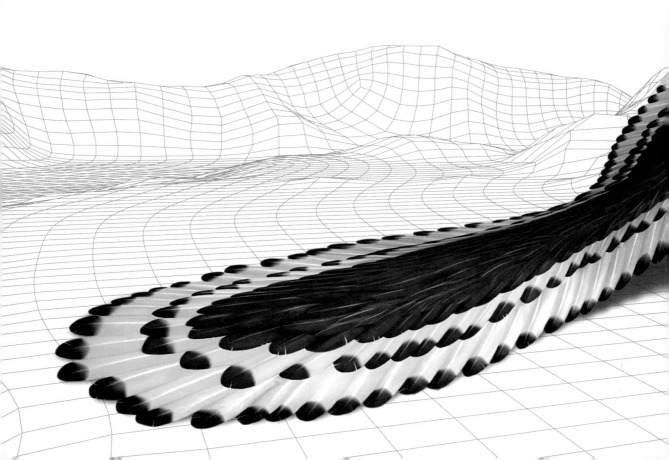

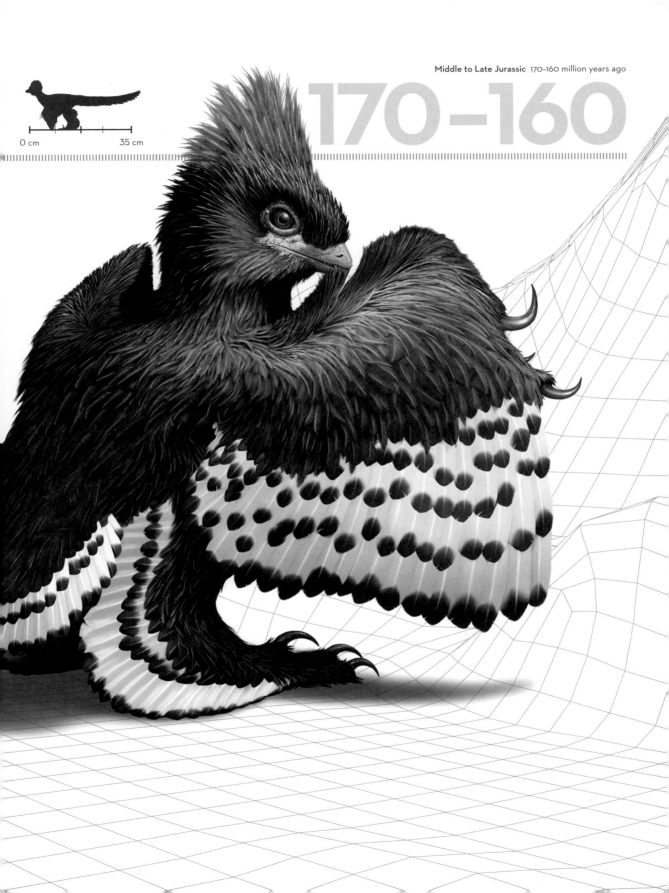

0 cm   35 cm

Speckles and spangles

Leaving our stripy *Sinosauropteryx* (pp. 28–41) behind, we travel another 40 million years back in time to the Middle and Late Jurassic, the time of the Yanliao fauna, an assemblage of plants and animals that lived 165 million years ago in northern China. Lush plants surround shallow pools, and all manner of fat insects – beetles, cicadas, dragonflies and cockroaches – scuttle about in the leaf litter. Salamanders swim in the pools and lizards laze around the water's edge, both happily feeding on the insects. The ferns by the waterside part and a beautiful black, white and red feathered animal rushes by.

This is *Anchiornis*, a small, bird-like dinosaur with huge wings and a long, feathered tail, as well as a red crest of stiff feathers rising from its head. Unlike *Sinosauropteryx*, *Anchiornis* has a wide range of feather types, not just simple filaments or bristles, but also fluffy down-like feathers on its breast, back and face, which mask the bases of the contour feathers on the wings, legs and tail. The feathers around the face are mainly black, with white and ginger speckles and spangles. But most striking are the contour feathers. As he spreads his wings, we see stripes of colour, made from these numerous banded feathers, which are white with black tips. The tips line up neatly, arranged like the tiles on a roof, covering the follicles and quills and giving the appearance of black stripes running right along the width of the wing and round the tail.

Contour feathers are the classic feather we pick up off the beach or road. They have a central quill – or properly the rachis – which is hollow. The rachis inserts deep into the skin in a follicle, a specialized pit, just like one of our hair follicles but with muscles and nerves, so birds can raise and lower their contour feathers and even rustle them for effect.

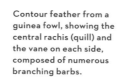

Contour feather from a guinea fowl, showing the central rachis (quill) and the vane on each side, composed of numerous branching barbs.

Opposite:
A modern zebra finch, showing feather colours all based on melanin. The black, grey, brown and blond colours are all based on eumelanin, and the ginger patch on the cheek is coloured by phaeomelanin (see pp. 52–53).

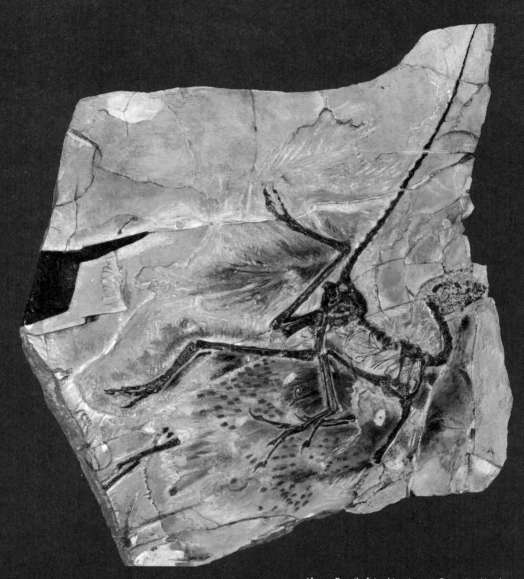

Above: Fossil of *Anchiornis* from Liaoning Province, north-east China, showing the legs extending left and the tail and head right. The wing feathers show remarkable detail of the mottled pattern from black-coloured ends on the primary and secondary flight feathers.

Opposite: Up close; an intimate view of a raven contour feather, showing the rachis sloping across the middle, the barbs branching upwards, and, between them, the barbules branching and interlocking. This exquisite structure enables the remarkably lightweight feather to provide an impenetrable surface for flight.

These contour feathers carry barbs on either side of the quill, and the barbs themselves carry side branches called barbules. If you run your fingers through the barbs of a pigeon or seagull feather, you can unzip them one way and zip them the other. This zipping and unzipping effect is because the barbs and barbules carry tiny hooks that interlock and turn the branching, frond-like feather into a single surface, essential of course when the feathers are used in flight. If the feather broke up, the air would rush through the wing and the bird would plummet from the sky. Birds today spend endless time preening, that is, running their beak through their contour feathers to remove twigs and seeds and to repair any unzipped portions. We can picture our proud *Anchiornis* doing just the same.

This *Anchiornis* is looking for a mate. He stands perfectly still, rustling his feathers as he stares into the distance. Then he skips from the ground onto a low branch in a monkey puzzle tree, and spreads his wings to full width. With wings outstretched and tail fully extended behind, he floats down to the ground: a spectacular show. The *Anchiornis* rustles his wing feathers expectantly, but nobody sees – his only audience is a fat beetle, and it is not impressed.

Melanin gives colour and pattern

The same patterning that we see in *Anchiornis* – a stripy effect achieved when the black tips of overlapping, otherwise white contour feathers line up – is commonly seen in modern birds. Seabirds in particular, such as terns, gulls and egrets, often have primarily white feathers with black tips. This is in part for physical protection: the black melanin pigment actually toughens the feathers, while there is no pigment at all in the white parts. This means that if modern seabirds brush the edge of a rocky cliff with the tip of their wing, the feather may be damaged a little, but less than if melanin was absent. Their predominantly white colouring serves a different protective purpose: camouflage. It allows them to blend in with the light reflected off the sea water. If a predator such as an eagle is soaring above, it looks for a shadow moving over the bright sea, and white birds are invisible. The black feather tip is an evolutionary compromise – too much black and the eagle spots the gull; too little and the tips of the feathers get damaged during flight in tight spots.

In fact, feathers and hairs are fundamentally transparent, although we see this as white. This is because hairs and feathers are made of the protein keratin, as are our toenails and fingernails. The nails contain no pigment, so the colour

of the skin shows through the transparent keratin, and it's the same with hairs and feathers. As humans, and other mammals, get older, they lose pigment from the hair, which goes first grey and then white or transparent.

It's different in *Anchiornis* because it was a dinosaur, not a seabird, and so it had no need for the white camouflaging effect. But evidently the ability of contour feathers to be white with black tips has a distant ancestry, and probably reflects a fundamental part of the genetic code of dinosaurs and their descendants, birds.

It may seem that animals, especially birds and insects, have a bewildering array of possible colour patterns, but this is not the case. Birds do not show every possible artistic arrangement of spots and stripes, as if produced by the paint splatters of an inner Jackson Pollock. The colours themselves are generally limited to blacks, greys, browns and gingers, especially in birds, for which melanin is the main pigment. But the range of patterns is also limited. In fact, feather patterns are very regular, and in the case of *Anchiornis*, what looks elaborate is in fact very simple – the spangled effect of black spots derives from the regular overlapping construction of the feathers, essential for the mechanical efficiency of the wing, coupled with a simple genetic command during feather development: 'send black melanin to the feather tip only'. Geneticists have found that particular developmental genes predetermine the palette of possibilities.

The seagull is primarily white in colour, so it looks flat or invisible to any prey animals, such as fishes, looking upward towards the sun. But the wingtips are black, caused by a concentration of eumelanosomes (see p. 52) in the feather tips, as a defence against feather damage.

Telling the colour of fossil feathers

How can palaeontologists be so sure they are establishing these colours and patterns accurately? The method was developed by Jakob Vinther, then a postgraduate student at Yale University, and now on the staff at the University of Bristol. He was interested in the exceptional preservation of fossils and was studying fossil squid and eye spots in fossils (see p. 188). The squid ink and the eye spots were always black, and he wondered whether this was because some trace of the original melanin was preserved.

On close examination under the Scanning Electron Microscope (SEM), he noted masses of sausage-shaped structures in the black areas. Previous scientific papers had identified these as bacteria; indeed, so-called coccoid bacteria are just this size and shape, and it was known that in many cases of exceptional preservation of soft tissues in fossils, such bacteria had quickly formed a gloopy film that created a microenvironment where oxygen was absent, and it was this process that meant the guts or muscles would not decay so quickly.

He then looked at an exceptionally preserved feather from the Early Cretaceous Crato Formation of Brazil. The fossil specimen showed clear dark and pale stripes across the feather, and the stripes were regular and V-shaped, as would be expected if these showed some hint of an original, biological pattern. If it had been caused by later damage, the stripes might have been irregular, or cut across feather barbs. He took small surface samples and examined them under the SEM: the dark areas were stuffed full of sausage-shaped bacteria, and the pale areas showed nothing at all – just rock. He looked at more samples, and found the dark areas were always full of bacteria, the pale areas not.

He took his idea to Derek Briggs, his supervisor, and a leading expert on exceptional preservation of fossils. At first Briggs was sceptical, but the evidence from the Brazilian feather was a clincher: 'There is no reason why bacteria would colonize one part of a feather and not another', he said, when they published these results in 2008. This was a revelation, causing many investigators to re-examine their fossil specimens, realizing the 'decay bacteria' they had identified might, in fact, be melanosomes.

The feather that started it all. A fossil contour feather from the Crato Formation, Early Cretaceous of Brazil (Leicester University, UK, Geology Department, LEIUG 115562) showing colour bands. The margins of the colour bands are similar to those found in living birds and not the result of damage. On closer inspection, the dark bands were packed with elongate 'bacteria', whereas the pale bands contained none at all; proof that the so-called 'bacteria' were really melanosomes.

Melanin and melanosomes

Melanin is a biological polymer that occurs in the skin, hair and feathers of vertebrates, as well as in the back of the eyeball, certain regions of the brain and some glands. Melanins occur in insect cuticle and in the eyes and eyespots of most animals. Melanins also produce the black coloration in the ink of the squid and other cephalopods, and they act as protective agents in certain microscopic organisms. In humans, melanin is best known as the colouring agent that makes our skin black, brown, reddish or yellowish.

There are several forms of melanin, which all differ chemically and produce subtly different colours. The one everyone thinks of is eumelanin. This provides primarily black colours, but also browns, greys and yellows, accounting for most of the hair and skin colours in humans and mammals. Blond and grey colours occur when there is less of the pigment. The other commonly found form is phaeomelanin, which produces reddish colours in the skin and hair. People with ginger-coloured hair have this unique pigment, and this is why their skin can be redder than in people who lack phaeomelanin. Phaeomelanin also makes red squirrels, foxes and some other ginger-coloured mammals red.

Modern birds of course show other colours such as greens and purples, and these come from pigments they consume, such as carotenoids (giving red and pink colours) and porphyrins (green and purple colours). Many people have seen the rather washed-out flamingos in zoos; they are perfectly healthy, but they are not getting enough carotenoids in their diet, a pigment that in the wild they get from the red shrimp they eat.

In the skin, melanins occur between the layers, and the amount present can increase or decrease and so change the skin colour, as for example when pale-skinned people become sun-burned. But in hairs and feathers, the melanins do not simply sit uncovered within the structure. Probably because hairs and feathers are composed of the protein keratin, which is flexible but a bit like plastic in texture, the melanin has to be inserted through a special process that occurs before the hair or feather sprouts from its follicle.

In these cases, the two kinds of melanin are synthesized in specialized cells in the skin, and these migrate into the base of a developing hair or feather in its follicle. There the melanin is packaged into melanosomes, and these melanosomes are transferred into the developing hairs or feathers. As the hair or feather grows, the melanosomes either occupy the whole structure, or migrate to particular areas.

Sometimes a single feather, for example from a zebra finch, may have multiple colour patches, perhaps ginger at the tip, black or grey lower down, and white at the base. This precise distribution of colour patterning is mapped by the presence and absence of the melanosomes containing eumelanin and phaeomelanin. These are genetically coded to be directed to particular parts of the feather, so that when all the feathers grow in their correct places, the zebra finch has a very exact ginger spot on its cheek, definite areas of black, grey and white, a brown and white spangled area over its flanks, and black and white spangles down its tail feathers. Just as in *Anchiornis*, the spangles in the zebra finch's tail are white tail feathers tipped by black.

Anchiornis preens her feathers. This very bird-like behaviour must have begun long before the origin of birds, among the theropod dinosaur ancestors. If you have feathers, they have to be preened, otherwise they would become tangled and sticky as you moved about in the trees and forest bottom. Some mammals lick their fur for similar reasons (think of cats), but not all (think of dogs). Feathers need preening, whereas hair does not, because feathers typically branch, and the barbs and barbules easily become unlocked and have to be teased back into shape.

Melanosome shape and colour

The matching of melanin type to colour and the fact that the melanosomes for eumelanin and for phaeomelanin are different shapes provides the key to determining the colour of fossil feathers. The first essential step was when Jakob Vinther showed that melanosomes could survive in fossils. And indeed, it turns out that ancient melanosomes are ridiculously common – nearly every fossil feather contains them.

This, in a strange way, has provoked scepticism among some commentators, who think 'this is too easy; they must be bacteria'. However, the fact is they are not. Many detailed studies have shown that Jakob Vinther was absolutely right in 2008: that these tiny 'bacteria' are locked deep into the keratin of the feather (rather than found on the surface), and that they are restricted purely to dark patches and absent in pale-coloured patches, is clear proof.

The secret is that eumelanosomes are sausage-shaped and phaemelanosomes are spherical. Even after millions of years, they may shrink a bit, but they keep their shapes. The detailed patterns seen in *Anchiornis* are based on close study of feathers from all over the body, and the density and shapes of the melanosomes reveal the colours. QED.

The mottled wingtip of *Anchiornis* is something of an optical illusion. On close inspection, it can be seen that this pattern arises from the regular arrangement of the flight and contour feathers coupled with black feather tips. In terms of the developmental control of colour, black wing tips are easier to explain than some form of speckling across the feather. This mode also ensures the speckles line up regularly, and the dinosaurs surely used the sharpness of the pattern in display, possibly to attract mates, just as birds do today.

These are the classic images taken by Jakob Vinther and published in his 2008 paper. He was studying the distribution of supposed 'bacteria' across the famous Crato feather (see page 49), small, sausage-shaped structures, each about 1 micron long – one-millionth of a metre, or one-thousandth of a millimetre.

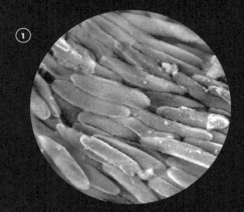

In the upper dark stripe (1), the microscopic elongate structures were equal in size and roughly aligned; the shape and alignment could indicate they were either bacteria or melanosomes.

In the first white stripe (2), Vinther found no trace of the elongate bodies; all he saw was rock. This convinced him the so-called 'bacteria' were definitely melanosomes.

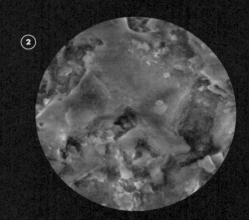

In the next dark band of the feather (3), the melanosomes occur in abundance, the same size and shape as in the other black stripe, and also aligned. If they were bacteria, they should occur all over the feather, as all areas, whether light or dark in colour, would be equally nutritious. He had the evidence that melanosomes survived abundantly and in good condition within fossil feathers.

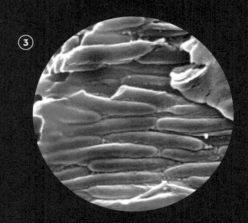

3

CAUDIPTERYX

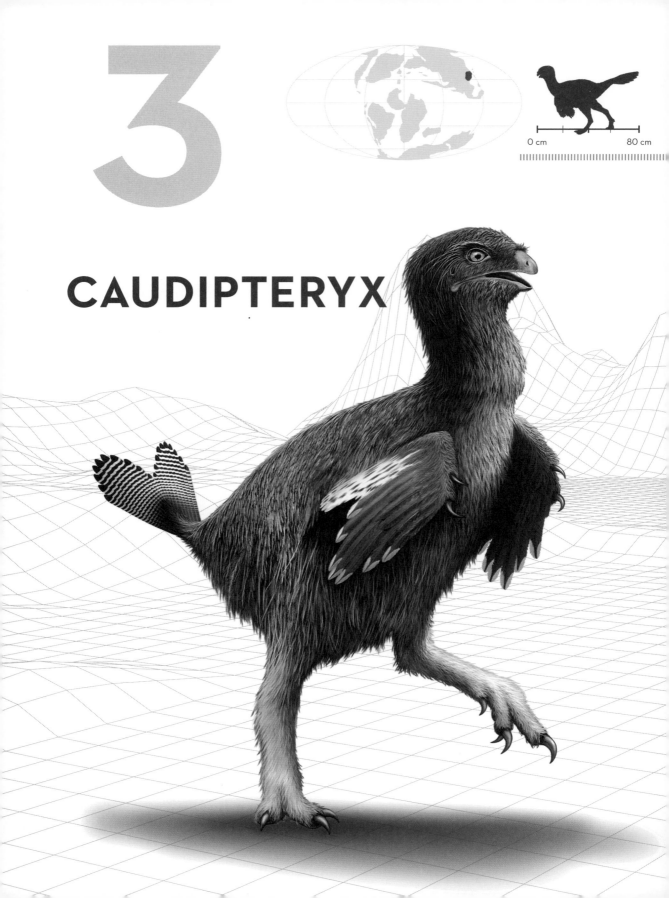

0 cm   80 cm

Flashing and shimmering

This is a most unusual-looking dinosaur, something like a grouse or blackcock at first glance… but its wings are tiny. This is *Caudipteryx*, a leggy and awkward-looking little dinosaur that lived at the same time and in the same locations as *Sinosauropteryx* (pp. 28–41). With its long, gangly legs, tiny, snub-nosed face, and tufty wings and tail, *Caudipteryx* seems even a little comical.

The long legs are partly covered with short white feathers in the lower half, and longer hair-like feathers over its powerful thighs. The rest of the body and head bear short, downy feathers, but great fans of feathers adorn the stumpy little wings and the tail. In fact, the wing feathers are much less extensive than in *Anchiornis* (see pp. 42–55), just attached along the second finger of the hand, and not on the arm at all. These are primary contour feathers (see p. 44). The tail fan is also limited – just at the tip of the bony tail, and arranged symmetrically, forming a heart-shaped tail.

We spot a male and female pair in action. The male on the right is dipping his head and spreading his stumpy wings; they flash white on and off as he extends and closes them out to his sides. He raises his tail, fanning it out as far as it will go. The barred brown and white feathers give a startling display, and as he raises the fan, he jitters the feathers so they shiver and shudder, giving an optical illusion of rapid movement accompanied by a sharp chattering sound. The blur of feathers is mesmerizing, but the female on the left has seen it all before. She squawks crossly.

As she opens her mouth, it can be seen to be a largely toothless beak, with just a few small teeth near the front. *Sinosauropteryx* and *Anchiornis* have many small, sharp teeth, and it is easy to see how they could feed on insects, lizards, and other small prey. These two aren't feeding, and it might seem to difficult to divine their diets.

Egg thieves, or not?

Caudipteryx was an oviraptorosaur. This group of rather unusual dinosaurs lived in the Cretaceous period, especially in what is now Asia and North America. All oviraptorosaurs have snub noses and a beak, and either tiny teeth, like *Caudipteryx*, or no teeth at all. In fact, *Caudipteryx* was the smallest oviraptorosaur, no larger than a turkey, but its relative *Gigantoraptor* from Mongolia was 8 metres (26 feet) long and weighed over a tonne. The skull and jaws give hints about their diet, but this has been a source of contention.

The first oviraptorosaur to be named, which in turn gave its name to the group, was *Oviraptor*. Complete *Oviraptor* skeletons were found by Roy Chapman Andrews on the first expedition of the American Museum of Natural History (AMNH) to Mongolia, in 1923. The trip was financed by the museum, and set off in a blaze of glory: Chapman Andrews had a flair for publicity, and sent back photographs of their heroic train of twenty black Model-T Ford cars as they embarked across the Gobi Desert in search of the origins of humanity. In fact they found no human fossils, but plenty of dinosaurs.

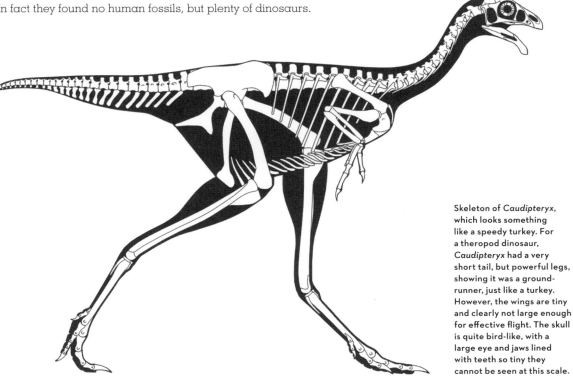

Skeleton of *Caudipteryx*, which looks something like a speedy turkey. For a theropod dinosaur, *Caudipteryx* had a very short tail, but powerful legs, showing it was a ground-runner, just like a turkey. However, the wings are tiny and clearly not large enough for effective flight. The skull is quite bird-like, with a large eye and jaws lined with teeth so tiny they cannot be seen at this scale.

The *Oviraptor* fossils were found close to skeletons of the small plant-eating dinosaur *Protoceratops*, as well as a nest of eggs. Could this be a case of predator and prey, the diminutive mother *Protoceratops* heroically defending her nest against the stalking, slender-limbed, beady-eyed *Oviraptor*?

Oviraptor featured prominently in the first exhibits of these amazing new dinosaurs at the AMNH. The Museum Director, Henry Fairfield Osborn, shared Chapman Andrews's knack for stirring up the media, as well as being himself wealthy and knowing many of the wealthy New Yorkers who funded the expeditions. In these fossils he saw the bones of a great story, and he had a tableau of the skeletons reconstructed, backed by a painting by the famous palaeoartist Charles Knight (see pp. 15–17). The excited citizens could see the drama unfolding in front of them; a humble, plant-eating Protoceratops standing guard beside her nest, a scraping in the ground that contained twenty or so eggs arranged in neat circles. The menacing skeleton of *Oviraptor* stood over the nest, claws out – ready to snatch. The fact *Oviraptor* had no teeth, just a sharp-edged beak, fitted the story: here was a predator that did not tear flesh with its teeth, but snapped at eggs and much smaller prey.

Chapman Andrews delivered the crates of bones to the AMNH in 1924 and Osborn had staged the exhibition by the end of that year, rushing out the scientific descriptions in a space of months. Osborn chose to call the dinosaur *Oviraptor*: 'egg thief'. It was the perfect story of heroes and villains, and a great draw for the public, who flocked to the AMNH to see these wonders.

However, the whole story was wrong. The nest and eggs belonged to *Oviraptor*, as was shown by a later AMNH expedition to Mongolia, in the 1990s, led by Mark Norell. Norell and his team found a similar nest, with identical eggs arranged in neat circles – and an adult *Oviraptor* sitting on them. This adult had shuffled into place carefully, so as not to break the eggs, and extended her fluffy wings out over the eggs near the edge. To clinch it, Norell and his team CT scanned the eggs, and identified the bones of embryo *Oviraptor* inside.

So, *Oviraptor* was not an 'egg thief', or at least not based on the evidence Osborn presented. Do we change all those names – *Oviraptor*, oviraptorosaur, oviraptorid? In fact, we can't. The rules that govern the naming of species are quite strict about priority – that is, the first name accepted by the scientific community. And, in any case, perhaps oviraptorosaurs like *Caudipteryx* were egg-eaters, or at least omnivores, meaning they might have eaten lizards, beetles and other small prey, as well as eggs, nuts and fruit, for example.

But what about those strange little wings? Surely *Caudipteryx* couldn't fly?

A classic museum display from the 1920s (1), evidence of the success of the American Museum of Natural History expedition to Mongolia: the skeleton of a mother *Protoceratops* standing guard over her nest.

The remarkably good preservation of fossil nests in the Gobi Desert of Mongolia (2), here showing nine rather complete, elongate eggs, plus scraps of many more, evidence that the mothers laid between fifteen and twenty eggs, arranged in a circular fashion, with the narrow ends pointing inwards to prevent too much rolling around. A later AMNH expedition showed that the eggs did not belong to *Protoceratops*, but in fact to the theropod *Oviraptor*, forever damned by its name as 'egg thief'.

Wings for flying and for flapping

Why would an animal have wings if it could not fly? When we think of wings, typically we think of creatures that flap them up and down, like most birds, bats and insects. But there are other kinds of flight. Biologists identify two other modes: gliding and parachuting. Gliders fly more or less horizontally, wings outstretched but not flapping, whereas parachuters plummet downwards, but slow the rate of descent with their wings. In either case, the animal doesn't flap its wings, but it does have them, and it can fly.

Examples include flying squirrels, gliding lizards, and some species of snakes and frogs. These all create a wing by adaptation of an arm, foot or even expanded ribs, and the 'wing' gives sufficient support to allow them to extend their leaps from tree to tree. Even if the leap is only 10 or 20 per cent further than it might have achieved without a wing, that could be enough for a gliding frog or lizard to escape a predator. Flying fish have elongated pectoral fins, and they use them to skim above the water, leaping as far as 50 metres (164 feet) at a time to escape predatory sharks.

To some, calling gliding flying might seem like playing with words, but for biologists the distinction between active flyers and passive flyers is a useful one. Flapping wings takes a huge effort, and so far, despite many attempts – some of them quite comical – human beings, with all their ingenuity, have entirely failed to make a flapping-wing aircraft. Pretty well all human-designed planes and drones use propellors and motors, and the wings are there simply to provide lift.

Lift is the key. In aeronautical terms, lift is the force that is opposite to gravity. In fact, lift is what enables birds and planes to defy gravity. Providing that lift is greater than gravity, the object goes up. In a rocket, the rapid combustion of fuel provides a huge down-blast of a force that exceeds gravity, which propels the rocket upwards. Forward motion can also generate lift – think of an aircraft racing down the runway, or a bird taking a few running steps before it leaps off the ground. Both plane and bird wings have a high, rounded leading edge that tapers to a thin, trailing edge. As the plane or bird moves forward, air splits over and under the wing, and because it has to travel further over the top, it goes faster, reducing pressure and 'sucking' the plane or bird up. That's lift.

But *Caudipteryx* would fail as a flyer, because its wings were too small. They are in fact so small that they would barely provide any support even for short gliding leaps from tree to tree. If it could not fly or glide, what on earth were its wing and tail tufts for?

Opposite:
The modern gliding lizard *Draco*, launching itself from branch to branch. Its 'wings', formed from skin stretched over expanded ribs down its side, enable it to extend the length of its jumps by ten times or more. With rib-based wings, it is unlikely this lizard could ever become a powered flyer.

Following pages:
The original specimen of *Caudipteryx zoui* from the Jehol Group of Liaoning Province, north-east China, now on display at the Geological Museum of China in Beijing. There are feather impressions behind the arms, preserved as black, carbon-rich impressions. In the gut region, a mass of stomach stones, perhaps used in grinding up plant food, suggests that this theropod might have fed on a mixed diet of flesh and ferns.

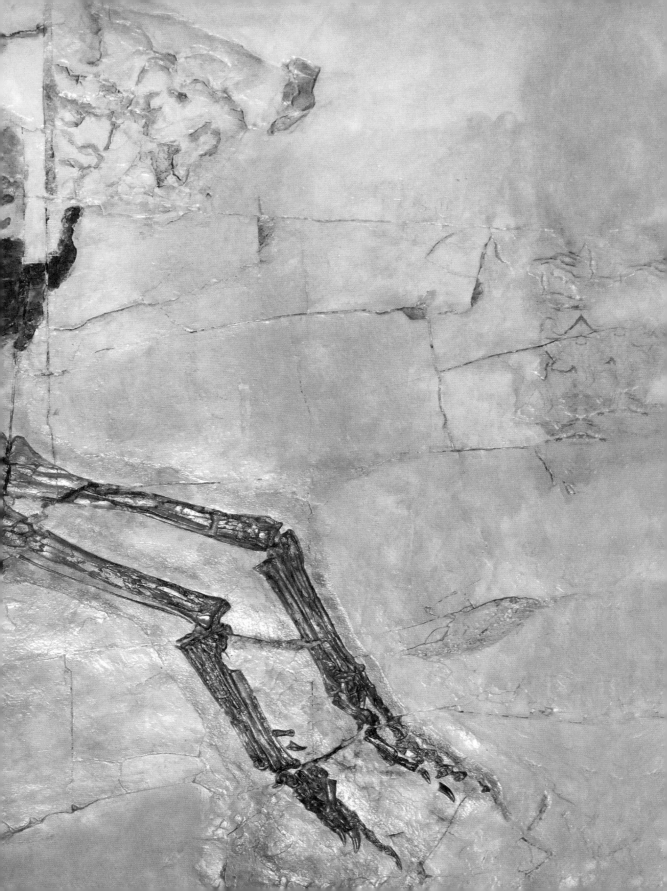

Wings for display

The original *Caudipteryx* specimen is kept in the National Geological Museum of China. It was described in 1998 by Qiang Ji and Shuan Ji, as well as Phil Currie and Mark Norell, researchers who had already studied early specimens of *Sinosauropteryx* (see pp. 28–41). They were the first to note that *Caudipteryx*'s contour feathers are symmetrical, which is often taken as evidence that they were not used for flight, based on comparison with modern birds: the wing feathers of a modern pigeon, for example, are asymmetrical, with the portions of the feather on either side of the rachis different in size, while the contour feathers of flightless birds, such as ostriches and emus, are more typically symmetrical.

The authors provided a detailed description of the skeleton of *Caudipteryx*, focusing on evidence it was an oviraptorosaur, and this is important of course in trying to understand something about the evolution of feathers in dinosaurs before they evolved into birds. They counted fourteen contour feathers attached to the hand and twenty-two to the tail, making a fan of eleven perfectly symmetrical feathers on each side. And while they could not comment on the colour of the feathers – it would still be twelve years before the method of determining dinosaur coloration was developed and applied to *Sinosauropteryx* and *Anchiornis* – they did include in their report colour photographs showing a bunch of five tail feathers, captioned: 'showing colour banding'. That was it; a rather understated approach to the first indication of pattern in dinosaur feathers, but back in 1998 there were still heated debates about whether these were really feathers or not, so the authors were probably being cautious.

This is work still in progress. Fiann Smithwick provides a detailed description of the plumage of *Caudipteryx* in his PhD thesis and it is yet to be published. But two of the fossil specimens show complete tail fans, composed of about twenty feathers, ten on each side. Each tail feather shows about fifteen colour bands, and these bands become narrower towards the tips of the feathers. Some specimens have longer tails with more prominent banding than the others, and these are possibly males. Their wings appear to show more white than the somewhat duller-coloured females.

Opposite:
The *Caudipteryx* wing feathers, showing distinct colour banding of these primary feathers. Notice the claws of the hand, one with a hint of a flipped-back keratin sheath that fitted over the claw bone below, evidence that the claws of the hand were even more fearsome than the bony remains at first suggest (see also page 34).

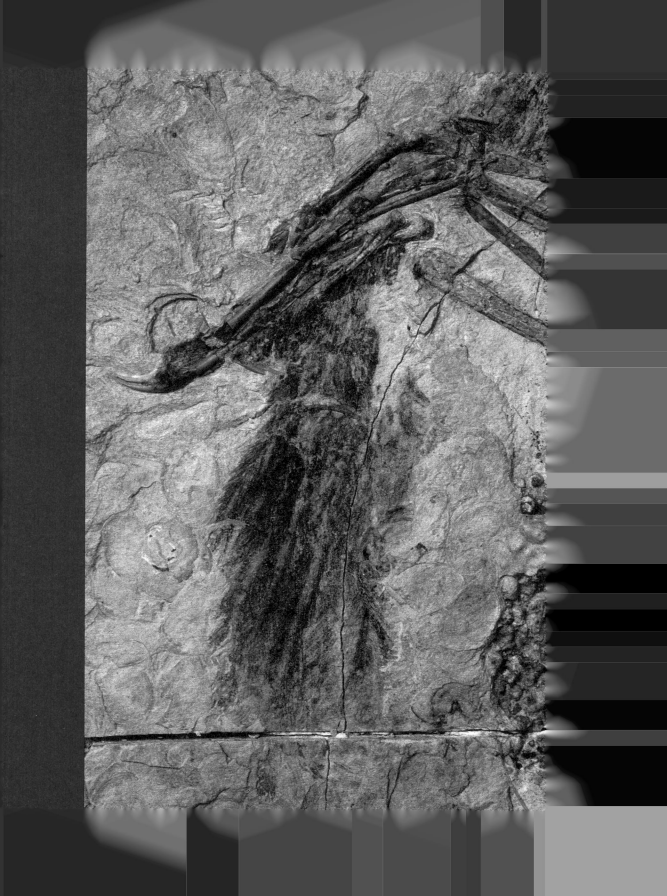

Colour banding on the *Caudipteryx* tail feathers. In this case, the bands appear to traverse each feather, a very different situation from the banding generated by wing-tip stripes on layered contour and flight feathers, as seen in *Anchiornis* (see page 54).

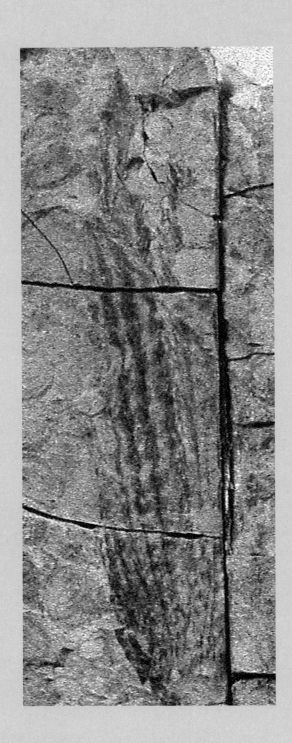

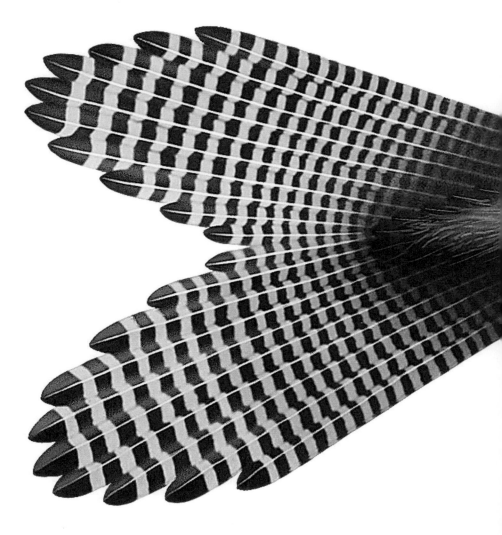

The *Caudipteryx* fan tail. This was not a flying dinosaur, but a road-running ground-dweller, and we might speculate that the striking barred tail was used for displays, as seen in certain game birds today, such as grouse and capercaillie. In lekkings, for example, groups of males pound up and down in a defined show ring, watched by females. The ritual displays avoid fighting and risk of injury, and the male with the largest feathers, or the most mesmerizing flashes and sweeps of the illusory tail bars, might win the right to mate with many females.

The colour patterns are for camouflage and for show: how can these be balanced? In fact, study of modern ground birds such as pheasants show how it works. At most times, both sexes keep their tails down and wings folded, so that they blend into the mixed brown and dull green of the vegetation in which they live. These are small animals, and there were plenty of predators.

However, at breeding time, the males in particular stretch their wings out, creating flashes of startling white. They lift their tails and rattle the feathers, giving an illusion of movement. Did they try to mesmerize the females, as male lyre birds do today, flashing sharp colours against the black of most of their feathers, and using movement and sound?

4

MICRORAPTOR

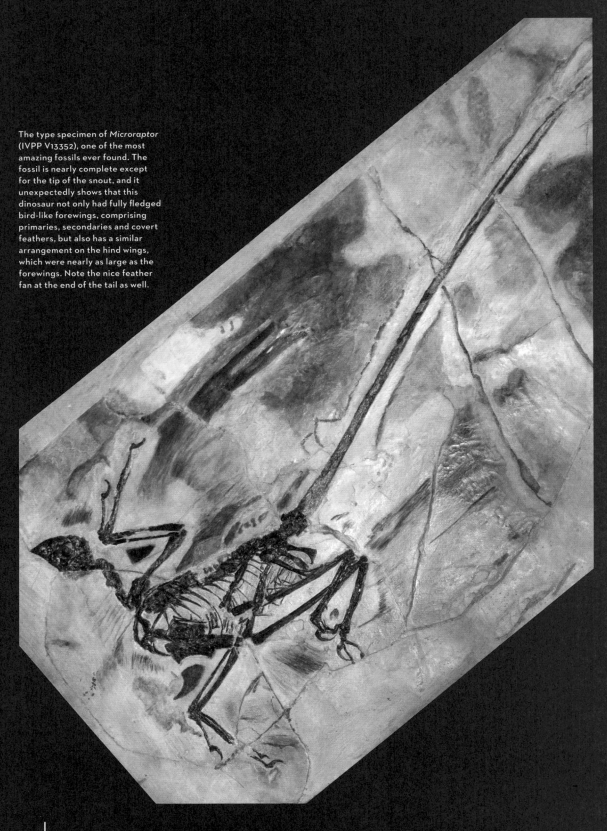

The type specimen of *Microraptor* (IVPP V13352), one of the most amazing fossils ever found. The fossil is nearly complete except for the tip of the snout, and it unexpectedly shows that this dinosaur not only had fully fledged bird-like forewings, comprising primaries, secondaries and covert feathers, but also has a similar arrangement on the hind wings, which were nearly as large as the forewings. Note the nice feather fan at the end of the tail as well.

The black hunter

The Jehol fossil beds of north-east China's Liaoning Province transport us back to the Early Cretaceous, some 125 million years ago. The climate is warm, but not tropical-hot; familiar-looking conifer trees stand on the higher ground, and horsetails and ferns line limpid pools. A fish breaks the glassy surface of the water, but misses a large dragonfly skimming just overhead, itself in search of smaller insects to eat. A *Sinosauropteryx* (pp. 28–41) marches past, waving its stripy ginger and white tail, and on the other side of the lake a small *Caudipteryx* (pp. 56–69) snatches at ground-dwelling insects.

Then, with a barely audible swooshing sound, a bright, shining, black bird-like animal leaps past. It lands neatly on the bark of a tree twenty metres away and chases up the trunk after a lizard. This is *Microraptor*. Not a bird, but a dinosaur. As it speeds up the tree, light falls on its strangely iridescent feathers, flashing steely blue, purple, black and startling white. The iridescence is like the blue of a kingfisher as it darts across a river seeking fish, the sunlight gleaming and blinking, and the blue brighter than any paint or print.

As *Microraptor* hurtles through the air, you notice it has four wings – two in front and two behind. No bird has ever adopted this kind of flying arrangement; it is a most unexpected discovery. In full glide, all four wings, two on the arms, two on the legs, flare out in an X-shape; and it can even beat those wings to keep aloft. As it comes in to land, *Microraptor* keeps the arm-wings fully outstretched in a glide, but tucks the leg-wings down to act as a kind of air-brake. Modern birds do just this as they slow down; by spreading the tail feathers and tucking them downwards, it's like an aircraft pushing out the flaps behind the wings to prevent stalling at slow speed. *Microraptor* lands precisely on target.

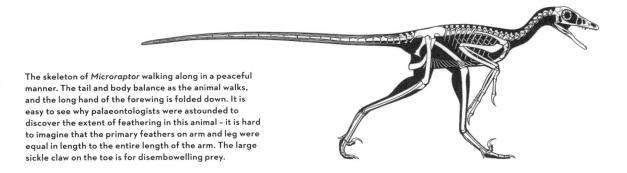

The skeleton of *Microraptor* walking along in a peaceful manner. The tail and body balance as the animal walks, and the long hand of the forewing is folded down. It is easy to see why palaeontologists were astounded to discover the extent of feathering in this animal – it is hard to imagine that the primary feathers on arm and leg were equal in length to the entire length of the arm. The large sickle claw on the toe is for disembowelling prey.

Dancing dinosaur.
This specimen of *Microraptor* lacks feathers, but shows every intimate detail of its skeleton. The mass of bones inside the rib cage belong to the lizard *Indrasaurus*. Indeed, this gut specimen is in fact the type specimen of *Indrasaurus wangi*. So, as it disappeared head-first down the gullet of *Microraptor*, the lizard could be encouraged by the fact that 125 million years later its remains would become a valued type specimen of an addition to the panoply of Cretaceous biodiversity.

Opposite: Melanosome types in a variety of modern birds, used to help calibrate the identification of likely original colours in *Microraptor* feathers. The Scanning Electron Microscopic images come from feathers that are brown (1), ('penguin-type') brown-black (2), grey (3), black (4), iridescent extant avian feathers (5), and *Microraptor* feathers (6). The samples come from titmouse, *Baeolophus bicolor* (1); macaroni penguin, *Eudyptes chrysolophus* (2); palm cockatoo, *Probosciger aterrimus* (3); Brazil duck, *Amazonetta brasiliensis* (4); double-crested cormorant, *Phalacrocorax auratus* (5); and the dinosaur *Microraptor gui*. The dots show the approximate location of sampling in each case. Courtesy of Quanguo Li of the Beijing Museum of Natural History.

The *Microraptor* hindwing, comprising a series of primaries, secondaries and covert feathers, just as in a modern bird forewing. The blue-black colour is hinted by the specimen (see page 72), where feathers seem all-dark, with no pale-coloured bands, but more importantly the closely aligned eumelanosomes (see pp. 54–55) inside the feathers indicate that the feathers were not only black in life, but also iridescent, reflecting light and flashing as the animal moved around.

Dining on the wing

Microraptor was named by Xing Xu in 2000, and since then three species have been identified, based on dozens of specimens. These are small animals, ranging in total length from 50 centimetres to 1 metre (1½ to 3¼ feet), but that includes the long, slender tail, and in life they probably weighed only about 1 kg (2.2 lbs) – the same as a plump chicken. The skull is short, but not particularly bird-like: its jaws are lined with numerous tiny sharp teeth.

In 2010, a *Microraptor* specimen was reported with tiny *Eomaia* (pp. 126–39) bones in its gut region. A year later, another *Microraptor* was described with bird bones in its gut, then in 2013 another one with fish scales, and finally in 2019 a specimen with a more or less complete lizard in its gut. It turned out that this unfortunate lizard was a species new to science, and it was named *Indrasaurus*. Evidently, palaeontologists, like *Microraptor*, are opportunistic in their feeding. This *Microraptor* had swallowed the lizard whole, and head-first; usually this would mean that it slipped down the throat smoothly and without thrashing about, but this lizard is in such good condition that it must have choked the dinosaur. He grabbed a prey animal that was just too big, and lacked the teeth and jaw muscles needed to break it up before swallowing.

How did *Microraptor* hunt? Descending quickly upon a tree was a great way to surprise its prey, even smart little animals like mammals, birds and lizards. It could even snatch fish from the lake below with its feet as it glided past. But the arboreal feeding is important. This perhaps gives a clue to the rise of the feathered dinosaurs in the Chinese Early Cretaceous. While their relatives were getting bigger and bigger, so they could dominate

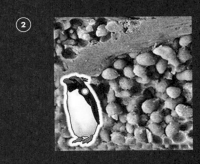

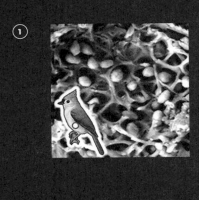

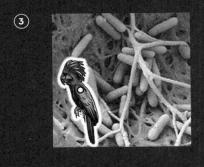

the ever-larger dinosaurian herbivores on the ground, a separate line of flesh-eating dinosaurs (theropods) were becoming small and feathered so they could concentrate on quite different prey.

The fossils from the Chinese Jurassic and Cretaceous periods show diverse faunas of insects, lizards, birds and mammals, all living around and in the trees. This was a rich food supply, and the feathered dinosaurs were at the top of the food chain. Being able to leap and glide gave them an edge in escaping larger predators, and in swooping rapidly onto other animals that were feeding on trees and in the foliage.

The *Microraptor* fossils with intact gut contents provide clear evidence that helps resolve a long-running debate about the origins of bird flight: was it from the ground up, or from the trees down? This debate began in the 1860s, as soon as the first bird, *Archaeopteryx*, was found (pp. 84–97), and *Microraptor* may resolve it. The lizard in its gut was likely a tree-dweller – so trees down.

Nanostructural colour

The unexpected iridescent black colour of all the feathers of *Microraptor* was determined in 2012 by Quanguo Li, Jakob Vinther, and many collaborators. They had expected to find patterns of stripes and spangles, as they had seen in *Anchiornis* (pp. 42–55), but instead found pure black – every feather contained only sausage-shaped eumelanosomes. There were no phaeomelanosomes that would provide any ginger tints, and no white patches or stripes suggesting that any feather lacked colour. And the investigators saw something else they had not expected, too.

The eumelanosomes in the *Microraptor* feathers lay parallel to each other, making up neat stacked sheets within the feather. By comparison with feathers from modern birds, including those of the common pigeon, the researchers found that this arrangement precisely matched that seen in iridescent regions of feathering. Such 'sheets' of aligned melanosomes are described technically as nanostructures because of their tiny size (a nanometre is one billionth of a metre, or one millionth of a millimetre) and this is the scale of many biological structures, as well as the functioning parts of computer chips.

Nanostructural colours are seen today commonly in insects and in birds. In beetles, for example, the carapace may be composed of multiple thin layers that are partially transparent and partially reflecting. There may be as many

Kingfisher diving. The flash of blue on its back is a reflection of the colour of the sky bouncing off its iridescent feathers. Technically, the kingfisher feathers are not iridescent, because the reflecting light does not bounce off a layered nanostructure, but they are semi-iridescent, reflecting light from a spongy structure of the keratin in the feathers. The flash of colour when a kingfisher dives may confuse its prey, which cannot quite see where the danger is coming from and so might be slower to react.

as ten such thin layers, each separated by a tiny distance, and as light strikes the back of the beetle it is partly reflected, partly absorbed, and so on down the stack of nanolayers. The effect is to turn a beetle with perhaps a dull green or yellow back into a living jewel: the colour is intensified many times over by the multiple reflections, which create an effect that is impossible to achieve with a pigment chemical alone. This effect, often called iridescence, is a mirror-like or metallic quality, and it can act as a warning signal or be deployed to confuse predators.

In birds, iridescence may sometimes have a camouflaging effect. For example, the kingfisher flashing along over the open water of a stream is made partly invisible if viewed by a predatory hawk from above. The intense flashing blue of its semi-iridescent feathers matches the sparkling of sunlight on the water below, although the kingfisher gets its blue from reflecting the blue sky above. The kingfisher can then dart from side to side in pursuit of a fish, while itself being reasonably safe from overhead predators.

In *Microraptor*, as in modern birds, the researchers explained, the iridescent colours were 'produced through coherent light scattering by laminar or crystal-like arrays generated by layers of materials with different refractive indices – namely, keratin, melanin, and sometimes air – in feather barbules.' So, the feathers had a base black colour, as we know from the

Following pages: Iridescent beetles, showing rainbow colours. Inside the beetle carapace, which is composed of two elytra, hard structures that protect the wings, are about ten layers of the polysaccharide chitin. In this beetle, the base colour is black from melanin under the carapace, and light from the sun strikes the multiple layers, partly being reflected and partly passing through to the next layer below. What we see is a complex of stepwise reflected light, giving either a single jewel-like intense colour or, as here, multiple rainbow-like effects. Study of these effects combines the skills of biologists and physicists; these natural nanostructures give inspiration for miniaturized computer-chip design.

abundant sausage-shaped eumelanosomes in all feathers. But the fact the melanosomes were lined up and packed in multiple layers gave them properties like the cuticle layers over the back of a beetle. Light penetrates a little and is reflected a little at each layer, and the end effect was the jewel-like, flashing colours reflected from dappled sunlight from the back and sides of the gliding *Microraptor*. Even round its mouth and face, just as in a modern pigeon, the colours would flash as it turned its head.

Modern pigeons are usually disregarded as vermin as they hop about scavenging for food in our big cities. But take a look next time you see a pigeon, and marvel at the iridescent greens and purples around its face and breast… The scrawny, limping street rat is actually at the pinnacle of avian evolution in its colour adaptations.

The biplane ace

Biplanes had a short existence in the history of aeroplanes. They were particularly successful in the early air forces of Germany, France and Britain during the First World War, when air aces like the Red Baron swooped and ducked over the battlefields of Flanders, taking photographs of the enemy trenches, dropping bombs and occasionally shooting at each other.

Biplanes have two pairs of nearly identical wings, one above the other, but with the lower pair set back a little. Using two pairs of wings meant that each could be reduced in size, as there is a simple aerodynamic principle that wing area is proportional to payload. A small plane has small wings, and a large plane has larger wings, but the relationship is between a square and a cubic measure: the weight of the plane is a cubic value, and wing area (a square value) has to be proportional. Doubling the length of a plane might correspond to an increase of four or eight times its mass, and thus the wing has to lengthen by four or eight times to carry the weight. Using two sets of wings, placing one above the other, roughly halves the required width of each wing.

But the air flows over each of the wings in a biplane interfere with each other and create extra drag, so there is a loss of effectiveness, and the wings have to be more firmly bolted together by firm struts to stop them shaking apart, which also increases drag. In the end, biplanes were more or less abandoned in aircraft design as faster speeds enabled monoplanes (those with a single wing) to limit wing size to some extent.

In nature, most flying animals have single sets of wings, one on each side of their body. Even dragonflies, which have two pairs of wings, don't arrange them above and below, but one behind the other. So it seems *Microraptor* was the nearest we can find to a biplane in nature. However, the two wing sets, on arms and legs, could alter their positions.

The two pairs of wings were very similar in construction and size, although the arms had a greater wingspan. Several investigations have been made of the likely flight performance of *Microraptor*, and the analysts had great fun making models from cardboard, wire and chewing gum, with chicken or pigeon feathers stuck in here and there. There is a long tradition of such model testing in palaeontology, and sadly many of the models have crashed.

This was not the case, however, for experiments by Gareth Dyke, Colin Palmer and colleagues at the University of Southampton. Their model was designed to allow them to change the spread of the wing feathers, retract or extend the legs, and vary the position and feather fanning of the tail. They placed the model in a wind tunnel – usually used to test engineering designs for drones – and ran it at different wind speeds and with the model posed in different ways. The wind tunnel allowed scientists to test a flying machine that is held steady, but the movement of wind mimics its different possible speeds.

It turned out *Microraptor* could glide in almost any posture of the wings and limbs, but performed best when the forewings and hindwings were horizontal and flat. In experiments where the model was made to glide from a perch, it achieved stability at high lift and high drag, and such conditions were best achieved when leaping from moderate heights of 20–30 metres (65–100 feet) and hitting the ground up to 100 metres (330 feet) away. If *Microraptor* held its legs out partly to the side it could cover 70 metres (230 feet) from a 30-metre (100-foot) high launch site, but if its legs dangled straight down, it could extend the distance it would glide to 100 metres (330 feet). Legs up or down – who knows?

New work published by Rui Pei, Mike Pittman and colleagues in late 2020 shows that *Microraptor* and *Anchiornis* were capable of powered flight… but only just. This was a surprising discovery, because these two animals are dinosaurs, not birds, and previously it was thought that the feathered non-birds could only glide. So, *Microraptor* was a pretty remarkable dinosaur, with some ability to flap and lift itself upwards, and capable of gliding however it launched itself, however the wings were held, and even in blustery winds that still would not have prevented it from staying aloft. An amazing animal in its day, but a failed experiment, as true birds emerged from the *Archaeopteryx* stock.

5

ARCHAEOPTERYX

0 cm   30 cm

The first bird

Archaeopteryx squawks, revealing her tiny teeth. She is only 15 centimetres (6 inches) tall. Her feathers are white, brown and black – black at the tips of her wings and fringing the tail, black and brown over the wing, and white round the neck and belly. The neck feathers look mottled, a pattern created by the dark tips of white feathers, as in her close relative, *Anchiornis* (pp. 42–55). She spreads her wings in a bid to cool down, after flapping her way to the ground. She is at home both on the ground and in the air. She has the general body shape of closely related dinosaurs, such as *Microraptor* (pp. 70–83) and *Anchiornis*, but has her main wings on her arms. The hind legs carry elaborate feathers, which functioned in flight. Likely, *Archaeopteryx* climbed trees, chasing insects, and leapt from branch to branch, much as *Microraptor* could. However, *Archaeopteryx* could flap its wings, and so it did not have to follow a downward trajectory. It could pump its wings a few times and go up – something the non-bird flying dinosaurs could not do.

Archaeopteryx lived in a very different world from that experienced by *Sinosauropteryx* (pp. 28–41), *Microraptor* and the other feathered dinosaurs and birds whose remains we now discover in modern-day China. Whereas we find those creatures fossilized in sediments deposited around ancient inland lakes, *Archaeopteryx* lived during the Late Jurassic, close by a warm shallow sea on the southern shores of an island continent that extended across what is now southern Germany. Indeed, its fossils are found in marine limestones that contain the remains of marine fishes, worms and sea urchins, as well as washed-in remains of plants, insects, dinosaurs, pterosaurs (flying reptiles) and birds. Likely the early flyers were blown off course by storms, and fell, exhausted, into the sea.

Opposite:
A classical quarry in the Solnhofen Lithographic Limestones in southern Germany. This 1889 lithograph was printed using the limestone that it shows being quarried. The limestone was quarried in slabs, and with careful application of the hammer and chisel split into sheets 1 centimetre (½ inch) or less thick. These sheets were perfectly flat, and they were used for printing after an artist had drawn the image using a wax pencil, which meant ink would 'stick' to the stone between the wax lines, but not on the wax, and so could transfer often complex images to printing paper. Here, the artist has produced a multi-colour image using multiple and different wax impressions.

The most famous fossil

At times, *Archaeopteryx* has been described as the most famous fossil in the world, the $10 million fossil, or the best proof of evolution. Indeed, the timing of its first discovery could not have been better. In 1859 Charles Darwin (1809–1882) published his book *On the Origin of Species*, the most important publication in the history of the discipline of biology. It documented for the first time together the two great principles of evolution: descent with modification and natural selection.

'Descent with modification', in Darwin's words, is what we might call 'the deep-time tree of life'. Darwin was the first to argue that all of life had evolved from a single common ancestor millions of years ago, and thus that all species today and in the past are related to each other, and we can best show the pattern in the form of a tree. Now it is possible to show the relationships of all 10,000 species of modern birds on a tree, based on analysis of shared patterns in their DNA. Fossils can be inserted into the tree based on evidence of shared characters of their skeleton and anatomy.

Darwin's other big advance in his *Origins of Species* was to provide the model for evolution in action through natural selection. He argued that individuals and species compete for resources such as food and shelter, and those organisms with traits that allow them to out-compete others tend to survive, and therefore have the opportunity to pass those favourable characteristics on to their offspring.

Opposite:
Maybe the most important single page from a notebook... Charles Darwin's famous sketch from late 1837, when he wrote 'I think' and drew an evolutionary tree. This was the moment when he realized all species were connected with each other in a single phylogeny, the closest species having split apart more recently, and distantly related species sharing common ancestors in much more ancient times.

I think

Case must be that one generation then should be as many living as now. To do this & to have many species in same genus (as is now). requires extinction.

Thus between A. & B. immens[e] gap of relation. C & B. the finest gradation, B & D rather greater distinction. Thus genera would be formed. — bearing relation

The first *Archaeopteryx* fossils were found in 1860 and 1861, hot on the heels of Darwin's book. What could have been better timing? Darwin's two big ideas about evolution had proved highly controversial, and his critics demanded proof. To demonstrate 'descent with modification' Darwin needed intermediate fossils, something say between a reptile and a bird, and here it was: the first *Archaeopteryx* fossil, now in the Natural History Museum in London. This remarkable find showed a complete skeleton with clear avian features in the feathers on its wings, body, hind legs and tail, as well as distinctly reptilian characters: teeth in the jaws and a long bony tail.

Darwin himself did not comment on the new finds; he was privately wealthy and by then a recluse, with no interest in fighting his corner by public debate in the scientific societies or in public. They became bones of contention, however, between the two leading anatomists of the day, Richard Owen (1804–1892) and Thomas Henry Huxley (1825–1895). Both had come from poor backgrounds and raised themselves by diligence and hard work, and had to seek university and government posts to provide an income. Owen had allied himself with the anti-Darwinians, Huxley with the new science of evolution, and both relished a public battle of intellect.

Each of them agitated to get their hands on the fossil. Through his higher offices, and authority as superintendent of the natural history department at the British Museum, in 1862 Owen was able to purchase the fossil, which had been found in a small quarry in Germany, for £700, worth a rather modest £86,000 today. He wrote the description, although the name had already been given a year earlier by the German palaeontologist Hermann von Meyer (1801–1869), and strove to show this was a fossil bird pure and simple, although much more ancient than anyone had ever expected for a bird. Owen argued that *Archaeopteryx* was as bird-like as any modern bird, and not in any way a 'missing link'.

Huxley, meanwhile, unable to see the fossil first-hand, nonetheless wrote about it, propounding his view that this was a dinosaur in birds' clothing – he compared its bony skeleton to that of *Compsognathus*, a terrestrial, predatory dinosaur with short arms that was also known from the same deposits in southern Germany. As we saw earlier (see p. 18), Huxley was on the right side of history, and his argument would be vindicated 100 years later by John Ostrom. Huxley had his theory published in newspapers and gave lectures to assemblies of scientists and of the general public, and so his view was widely dispersed and widely accepted.

Opposite:
The Berlin specimen of *Archaeopteryx*, collected from Solnhofen in 1874 or 1875, purchased from its finder, the farmer Jakob Niemeyer, for 20,000 gold marks by the Berlin Museum of Natural History, now the Humboldt Museum. This was the second specimen to be offered for sale; the first one had been purchased by the Natural History Museum in London, and so the Germans were determined to secure this one. Hence the very high price, which was financed by Ernst Siemens, founder of the Siemens company.

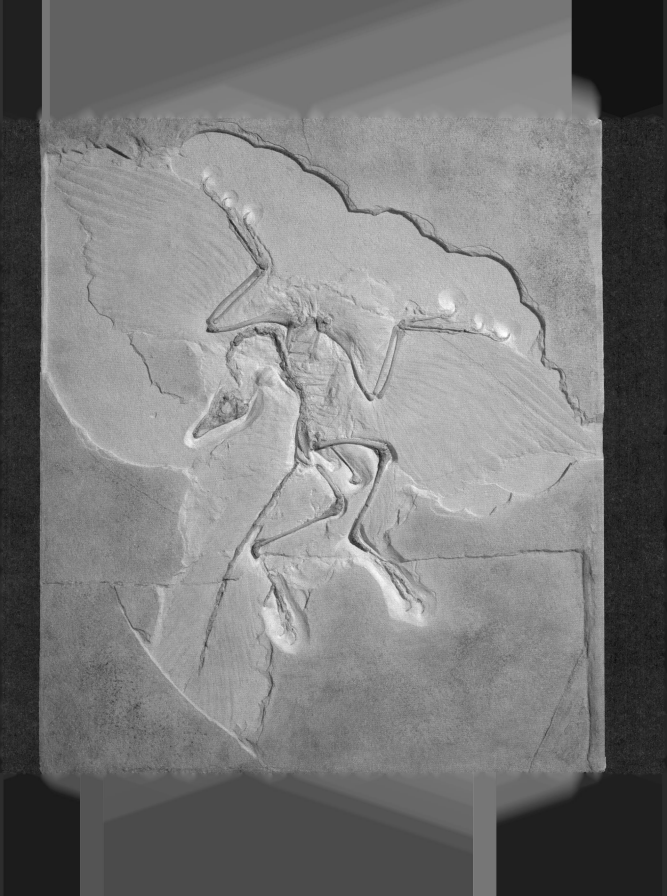

The 1860 feather: the first evidence that birds existed in the Mesozoic. This specimen was found before the *Archaeopteryx* skeletons and studied in great detail. Now, the main slab is in the Natural History Museum in Munich, and the counterslab in the Humboldt Museum in Berlin.

The feathers of *Archaeopteryx*

Over the next 100 years, more *Archaeopteryx* specimens were found in the south German quarries, and it became a favourite of textbooks and popular dinosaur books. Nobody yet had any idea how to extract the ancient colour from the fossil feathers; reconstructions often showed *Archaeopteryx* wonderfully coloured, cloaked in blues, greens and reds, but such speculation was entirely without evidence. The feathers themselves, by contrast, came under close scrutiny.

The wings of the 1861 or London specimen, as it is called, show all the flight feathers – the long primaries in the outer part of the wing, and the shorter secondaries and tertiaries closer to the body. The feathers are asymmetrical, just as in the wing feathers of modern flying birds, and unlike the symmetrical wing feathers of *Caudipteryx* (pp. 56–69). The covert feathers (vaned feathers that cover others) of the tail are more symmetrical, and arranged in series all the way down the long, bony tail. In modern birds, the long bony tail of their dinosaurian ancestors has shortened to a stump, and all of the tail feathers are attached to that stump like a kind of fan. But early birds like *Archaeopteryx* retain the dinosaurian arrangement, as we saw in *Microraptor*, where the tail feathers form more of a frond pattern, with feathers branching one after the other all along the length of the bony tail.

There were also some short covert feathers down the middle of the back, and the rest of the body was probably covered with short, whisker-like feathers, as in *Sinosauropteryx*, or fluffy down feathers. These are not as clearly seen as in the Chinese specimens, as the mode of preservation is different, consisting of impressions of the feathers with less organic material in them than in the Chinese volcanic lake deposits of Jehol and Yanliao; perhaps the smaller, fluffier feathers were lost before burial.

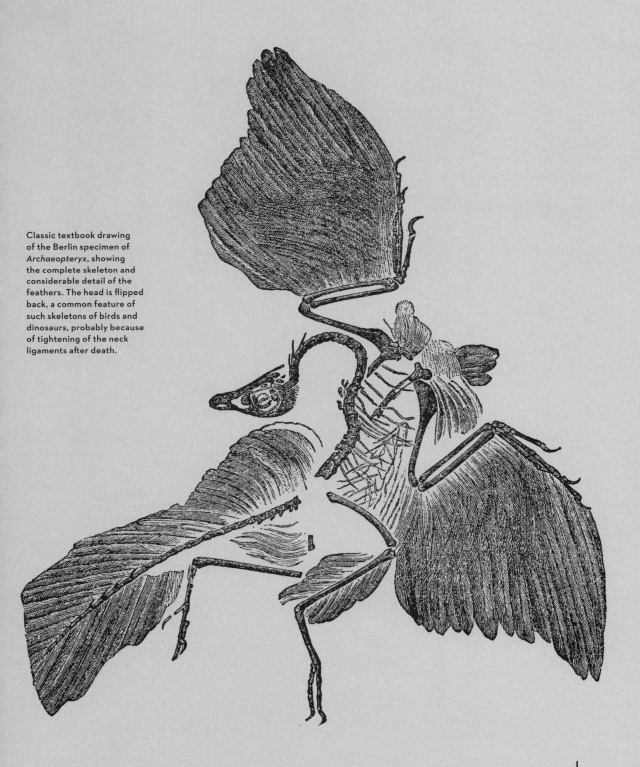

Classic textbook drawing of the Berlin specimen of *Archaeopteryx*, showing the complete skeleton and considerable detail of the feathers. The head is flipped back, a common feature of such skeletons of birds and dinosaurs, probably because of tightening of the neck ligaments after death.

A study of the colour of *Archaeopteryx* feathers was published in 2012 by Ryan Carney, then a graduate student at Yale, but he was allowed to analyse only the isolated 1860 feather, and no others. Its melanosomes showed that the feather had been black, with perhaps the darkest pigmentation near the tip of the feather. Unfortunately, there have been no further studies of the *Archaeopteryx* feathers.

Why just one feather? In order to do colour studies, it is necessary to take a small rock fragment from a fossil specimen to examine under the microscope. Even when we have explained that the samples we need are very small – just a millimetre or two across – museum curators nevertheless can go a little green around the gills, worried, quite understandably, about damage to their priceless specimens. In the case of the 'world's most valuable fossil', it's no wonder Carney could only study the single, isolated feather!

Could the first bird fly?

It might seem self-evident that if it looks like a bird, it behaved like a bird. *Archaeopteryx* has bird-like wings, and the shoulder joint looks like the shoulder joint of a modern bird, so it likely also had similar flight muscles. But the development of flapping flight constitutes a great evolutionary leap. Gliding is so much easier, and at first sight *Microraptor*, which certainly could not fly by flapping its wings, looks just as bird-like as *Archaeopteryx*.

So could *Archaeopteryx* fly? Well, it does not have the deep, keeled sternum bone seen in modern flying birds. This bone is familiar to anyone who has carved a chicken. We lay the chicken on its back and slice the white breast meat away from the sternum, which sticks up on the carcass in front of us. The breast meat, made primarily from the great pectoralis muscle, is the source of flapping power. The pectoralis runs from the top of the humerus (upper arm bone) and attaches all across the sternum, so that when it contracts the wing is pulled down and back with great power.

The wing upstroke in birds is essential, of course, but less significant. This is powered by the supracoracoideus muscle, which is tucked under the pectoralis. It runs from the back of the humerus through a special pulley over the shoulder and down to the sternum. When it contracts, the wing goes up, but that movement requires much less power than the downstroke, and so the supracoracoideus is maybe a tenth the size of the pectoralis. If you buy your chicken already carved, you will find that each chicken breast divides into unequal parts, the larger pectoralis, and the smaller supracoracoideus nestling in the side. (If you have a more gruesome turn of mind, you can pick up a dead pigeon, clean off the breast feathers and skin, and separate the pectoralis and supracoracoideus muscles. Then, pulling each in turn, the wing goes up and down. We discovered this as zoology students, and then fried the pigeon breasts – a delicious snack, eaten with toast.)

Archaeopteryx did not have this large sternum crest, just a low structure, and so some argued that it must have lacked a large pectoralis muscle and therefore must have had, at best, a weak downstroke. This is likely true, but bats are the same, and they seem to be able to fly by flapping their wings perfectly well.

In fact, calculations of the likely body mass of *Archaeopteryx* and its wing area show they are just right for flight. The wing was large enough to support the body mass, and the overall wing size and shape was that of birds such as crows and pheasants that fly through cluttered environments, full of bushes or trees. So, *Archaeopteryx* could fly – not for long distances, like a gull or tern, or for speed, like a swift or martin, but probably only for short flights of tens or hundreds of metres at best, weaving between trees and obstacles.

Trees down or ground up?

Experiments with a flying model of *Microraptor* (see pp. 70–83) showed that gliders performed best when leaping from a high point and controlling their rate of descent. The same is likely true of *Archaeopteryx*. Ever since the first fossils were found, ornithologists have debated heatedly whether birds evolved flight from the trees down or the ground up. The trees-down hypothesis prevailed until the 1970s, when John Ostrom argued for a ground-up theory – that feathered dinosaurs ran faster and faster, leaping up to catch insects with their feathered wings. Eventually, jumping and hopping, they took off, and flapped their wings to fly.

This might seem a little far-fetched, but the ground-up idea was supported in the 1980s by aerodynamics experts who argued that gliding and powered flight had evolved independently, and that a glider could never evolve into a flapper. Their main argument was that gliders lack the powerful muscles needed to flap a wing. In addition, we see no signs of modern gliding lizards,

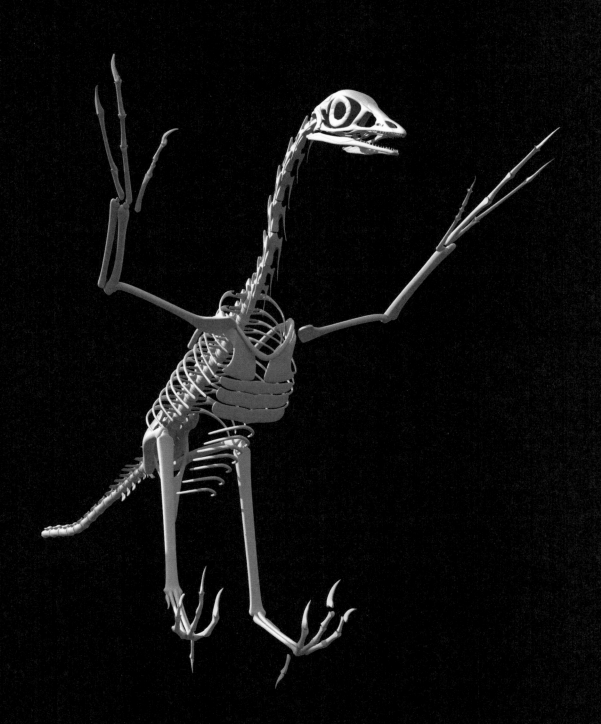

frogs and mammals becoming flappers. Further, in work by Kenneth Dial from the University of Montana, it is evident that some ground-dwelling birds, such as the chukar partridge, provide an interesting model for how a bird could start running up slopes before taking off into flight. His model of wing-assisted incline running is a very popular view.

However, all the new fossils from China point to a trees-down origin of flight. Gliders can in fact become active flyers, despite what had been said. Second, most experts in biomechanics would say 'use gravity, stupid'. Jumping down saves energy, and a plethora of gliders do just this today.

This impacts our understanding of *Archaeopteryx*, and all the other dinosaurs and early birds with wings. Whereas most theropod dinosaurs, such as *Allosaurus* and *Tyrannosaurus*, became larger and larger as they evolved to be ever more deadly hunters of the giant herbivorous dinosaurs, one evolutionary branch of flesh-eaters became small and sprouted wings. The evidence of their environments suggests wooded landscapes, and their diets seem to have been those of tree-dwellers, so chasing prey in the trees seems to explain the size reduction and initial forays into flight.

Opposite:
Archaeopteryx coming into land. Without the feathers, this model of the skeleton looks barely capable of flight, but of course the feathers on the wings and tail provide broad support surfaces that generated lift. But for *Archaeopteryx*, as for all other flyers, landing is a dangerous business. First, the bird has to slow down to a dangerously low speed, at which it risks stalling. The slow speed is essential to make sure the bird does not simply crash into the ground and break its bones. Very likely, *Archaeopteryx* spread out its wing and tail feathers to maximize the flight support at slow speed. The legs and feet are extended forward, and they have to absorb the impact of landing; coming in sideways to land on a tree trunk might be less dangerous than simply landing on the open ground, as the early bird could grasp the trunk with its foot claws and not risk toppling over.

6

CONFUCIUSORNIS

50 cm

Shades of grey

The male bird emerges from the trees, hopping along a branch, and landing
in a clearing on the ground. He makes a show of his long tail feathers,
squawking, preening and ducking his head up and down to attract the
attention of a nearby female. Like a peacock, he could probably raise and
shake those tail feathers, making a distinctive rustling sound. He puts a great
deal of energy into the show, while the female, short-tailed and slightly more
dull in colour, stands at the side, watching him – or perhaps looking into the
distance and thinking of lunch.

    This is *Confuciusornis*, the most common bird in China 125 million years
ago, in the Early Cretaceous. Thousands of *Confuciusornis* specimens have
been found, many with exquisitely preserved feathers, sometimes even
showing patterns of dark bars and speckles. Early attempts to reconstruct
their plumage hinted at gingers and other colours, but a comprehensive study
in 2018 led by Quanguo Li of the China University of Geosciences in Beijing
and Julia Clarke of the University of Texas revealed a spectrum of greys and
blacks. The wings were uniformly grey, except for a series of dark feather
tips under the wing. As these feathers overlapped, the dark tips formed five
discontinuous bars. The body was also mainly a grey colour, except for the
neck and head. The throat feathers are pale and marked by black speckles

and crescent-shaped dark patches, which continue round the eyes and cover the sides and top of the head, leading to a series of spiky feathers pointing back from the back of the head – very punk.

The long, banner-like tail feathers of the males have been of particular interest. The specimen studied by Li and Clarke was a female, but other work shows those long, slender feathers in males may have been grey along most of the length, perhaps with a pale grey stripe down the middle representing the uncoloured feather shaft, and with tufts of black at the ends. Li, Clarke and colleagues suggest that the feathers that lack melanin pigment over much of the body might have been coloured red, yellow, green or purple with pigments that cannot be detected. If so, the pattern would have been even more striking.

In modern birds, feather patterns and colours are used for camouflage and signalling. Their value in signalling is in providing information: sexual signalling, when males show off to females in an attempt to attract mating partners, is an obvious example, and core to evolutionary success. But feather colours and patterns can also inform other birds of the species you belong to, your sex, age, and condition of health. *Confuciusornis* had very elaborate patterns of stripes and speckles, as elaborate as in any living bird, and this shows how important the detailed coloration and patterning of feathers was even early in bird evolution. Their general shades and patterns also provide information about habitat and ecology.

Reconstruction of the grey speckles that have been discerned in *Confuciusornis* specimens. This is among the earliest evidence of elaborate feather patterns, which might have evolved for camouflage, or as a way for birds to signal their species or sex.

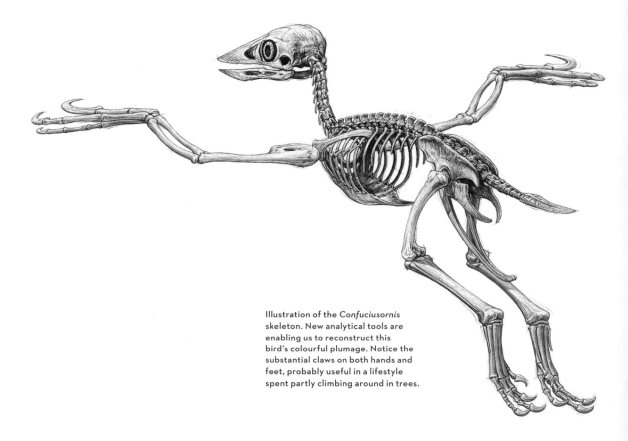

Illustration of the Confuciusornis skeleton. New analytical tools are enabling us to reconstruct this bird's colourful plumage. Notice the substantial claws on both hands and feet, probably useful in a lifestyle spent partly climbing around in trees.

Getting the colour out of the fossils

Li and Clarke's 2018 study is a great example of the range of cutting-edge methods now available to palaeobiologists exploring the colour of ancient birds and dinosaurs. First, the researchers took thirty-two tiny samples from a single fossil held in the China University of Geosciences collections. They then applied a battery of expensive analytical tools to explore the nature of the pigments: Scanning Electron Microscopy, Confocal Raman Spectroscopy, Time-of-Flight Secondary Ion Mass Spectrometry, and Matrix-Assisted Laser Desorption/Ionization Mass Spectrometry, used here for the first time in the study of ancient melanin.

The Scanning Electron Microscope (SEM) is a commonly used tool in all areas of natural science and is the first means of identifying ancient melanin, the colour pigment that resides in capsules within feathers called melanosomes. Under the SEM, minuscule fragments of the fossilized feathers can be enlarged 100,000 times. At this high scale, the melanosomes,

Confuciusornis specimens, one presumably male (left), one presumably female (right), preserved on a single slab of limestone from Sihetun, China. The male's long, banner-like tail feathers may have been used in sexual display.

Melanosomes in fossilized feathers. Whether the melanosomes are spherical (1) or cylindrical (2) indicates the colour of the pigment that they contain. The spherical phaeomelanosomes are typically ginger, and the cylindrical eumelanosomes blacks, browns and greys.

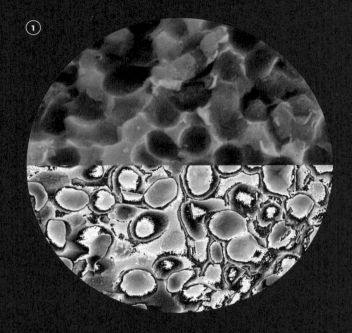

each measuring 1 micron – or one-thousanth of a millimetre – can be seen. The researchers identified melanosomes of all different shapes, from classic sausage-shaped eumelanosomes to spherical phaeomelanosomes, in the *Confuciusornis* specimen. These shapes indicate the original colours carried by the capsules: broadly blacks, browns and greys for the eumelanosomes, and ginger for the phaeomelanosomes.

But the researchers were not primarily interested in melanosome size or shape, but in whether they could demonstrate the presence of melanin in the fossils using chemical analytical techniques. They first employed Confocal Raman Spectroscopy, a method that allows the investigator to establish information about depth and buried structures in a sample by the use of a powerful laser beam. This laser beam is capable of focusing on and providing chemical analysis of a tiny spot, just 1 micron across. By comparing the spectrum of wavelengths encountered in the ancient specimens with the signals from modern, chemically pure melanin samples, Li, Clarke and colleagues were able to demonstrate the presence of eumelanin in the fossils.

The third mode of analysis they conducted on the fossil samples was Time-of-Flight Secondary Ion Mass Spectrometry (ToF-SIMS), using a million-dollar machine. ToF-SIMS, like Confocal Raman Spectroscopy, is capable of analysing tiny samples. It does this by impacting the sample with a pulsed particle beam that dislodges ions and hurls them towards a detector; their time of flight, measured in nanoseconds, can be used to identify the chemical compound.

The researchers tried applying other methods to discern the chemical composition of pigments, but sadly these did not provide conclusive results. That's not to say they are a waste of time – sometimes the pioneers who are first to apply a method struggle to get it to work, but it will be tried again, and a protocol might be figured out. Nobody would have predicted that such hi-tech analysis as this would be possible when the *Confuciusornis* specimens were first collected in the 1990s.

Reconstruction of the coloration of the *Confuciusornis* wing. Notice that the feathers have unzipped here and there, possibly as a result of his energetic dancing.

The Jehol Group

There are more than 3,000 specimens of *Confuciusornis*, the 'Confucius bird', in Chinese museums, and many of them come from a single locality, Sihetun, near Beipiao, in north-west Liaoning Province. The quarry is long and white, dug down in the middle of farmland, its walls 10 metres (33 feet) high. When we first visited, in 2007, the Institute of Vertebrate Paleontology and Paleoanthropology in Beijing had just carried out a large and productive excavation.

Typical specimens of *Confuciusornis* are squashed flat like roadkill, so they are seen when a pair of slabs is split apart. These slabs of limestone may be only 1 centimetre (½ inch) thick, but there is an exquisite matching fossil on each. Most examples are about the size of a pigeon. The wings are usually slightly unfolded, the legs spread apart, and the head bent over. In the middle, the flesh and internal organs have collapsed into a black organic mass below and around the bones, and spread out to the sides are the feathers.

While geologists had explored the fossil-bearing sediments in this area in the 1920s, it was another seventy years before these specimens were found, or at least published in the scientific literature. The early geologists identified a unique rock unit spanning a huge area, measuring thousands of kilometres from west to east across north China, particularly in Liaoning, Hebei and Inner Mongolia Provinces, and named it the Jehol Group. They noted rich assemblages of fossil insects and other invertebrates… but amazingly no birds, dinosaurs, or even isolated feathers were reported, until the site was revisited in the 1990s. Perhaps farmers had been turning *Confuciusornis* fossils up in the intervening years, but hadn't mentioned them to anyone.

So many of the most remarkable fossils in recent decades have been discovered in China, and in the case of dinosaurs and early birds, primarily from the Jehol Group, which consists of the Yixian and Jiufotang Formations. This poses a key question: why does one country produce so many astonishing fossils? One answer is that palaeontologists have been collecting actively throughout Europe for 200 years, and in North America for perhaps 150 years, and so we might not expect to find much that dramatically changes the fossil record in these regions. This isn't entirely true, of course, but palaeontology in China only really kicked off in the 1990s, and the country has proved to be a treasure-trove of amazing fossils. These are not all dinosaurs, but include much older rock units yielding some of the oldest animals, others with very early fishes, and yet others with amazing marine reptiles.

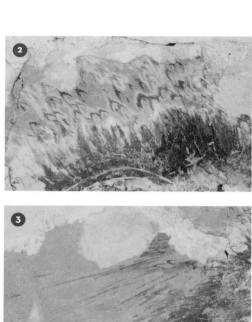

Pre-mating display

The most famous *Confuciusornis* specimen shows two individuals together, one with a short tail, and one with long tail feathers. These have always been interpreted as female and male respectively, the male adorned with banner-like tail feathers, as we saw earlier. We might think of modern peacocks, lyrebirds or pheasants, where the males are gorgeously brightly coloured, and use their tails to great effect in pre-mating displays. Was this the case in *Confuciusornis* too? Did it strut, and raise and lower its tail, perhaps rustling the feathers to draw attention to the display, as peacocks and turkeys do?

These behaviours are hypothesized, based on comparisons with modern birds. But can we even be sure that we have identified the males and females correctly? After all, we might simply be looking at two different species, one with a short tail and one with a long tail. Confirming the sex of these *Confuciusornis* specimens is still not certain.

The first effort was to explore the idea that only males had long tail feathers, and only females had short tail feathers, and that perhaps males and females would differ in size. But, in a 2008 study, Luis Chiappe from the Los Angeles County Museum and colleagues found that there was no clear correlation between body size and the presence or absence of long tail feathers in these birds. They measured 100 *Confuciusornis* specimens, identifying a juvenile and two adult size classes. Among the adults, one size class clustered around a body mass of 300 grammes (10½ ounces), the other around 500 grammes (17½ ounces). In modern birds, it is often the female that is the heavier adult, as they have to produce and lay eggs. But it was not clear which of these size classes of *Confuciusornis* was male and which was female, or even whether the two size classes represented the two sexes. The researchers found that long tail feathers occurred in specimens through the entire size range, from 150 to 700 grammes (5¼ to 24½ ounces), and the same is true for the absence of ornamental tail feathers. Perhaps, they suggested, males and females both grew long tail feathers at certain times of the year, something like pheasants today, where the females have tail feathers as long as those in the males, although less colourful.

Opposite: Evidence of plumage diversity in the Confuciusornithidae from specimen CUGB P1401. (1-4). The primary slab of CUGB P1401 showing details of the plumage including crest ornamentation on the (2) top and (3) back of the head as well as on the (4) secondary and (5) gular feathers (feathers that cover the throat region).

BODY SIZE AND TAIL FEATHER LENGTH

Luis Chiappe studied the correlation between the size of
Confuciusornis specimens and the presence of long tail
feathers to determine whether these feathers are indicative
of sex. It turns out not to be as simple as had been assumed –
birds with ornamental tail feathers (yellow bars and squares)
are present at all body sizes, matching the size ranges of
those without ornamental tail feathers (white bars and squares).
Long tail feathers are seen in both juveniles and adults
(size at sexual maturity is marked), so it's not clear whether
or not long tails are a male-only feature, but unusually then
present from a juvenile age.

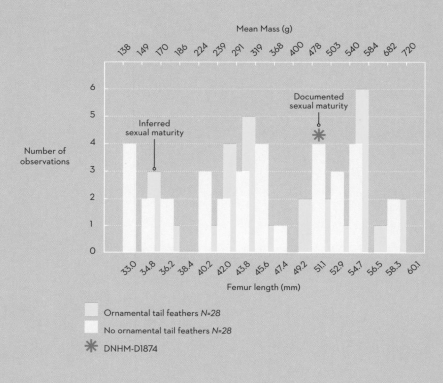

Ornamental tail feathers *N=28*

No ornamental tail feathers *N=28*

DNHM-D1874

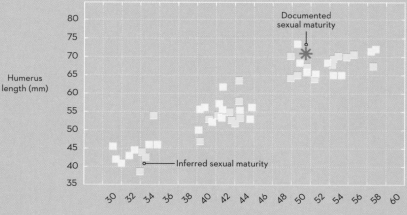

A few years later, in 2013, Anusuya Chinsamy from the South African Museum and colleagues confirmed that at least one *Confuciusornis* specimen with a short tail was female. They were examining thin sections cut through the arm and leg bones when they spotted some medullary bone. This is a specialized spongy bone tissue found only in female modern birds. It forms inside the medulla, or central cavity, of the bone just before the female bird is ready to lay its eggs; calcium, needed to generate the eggshell, is extracted rapidly, leaving a characteristic spongy, irregular bone matrix behind.

So, did *Confuciusornis* males behave like modern pheasants and peacocks, using their ornamental tail feathers in sexual displays? Even if females also had ornamental tail feathers, both sexes perhaps sprouting them only at certain times of the year as signals of health and strength used in selecting partners, this does not rule out that there was also some sexual differentiation of the kind observed in modern birds. More work is needed to determine whether some long tails were more brightly coloured than others; perhaps, like the modern pheasant, males had more colourful plumage than the females they hoped to attract.

Might a male *Confuciusornis* have used his colourful and extravagant tail feathers in sexual display, as the male peacock does?

7

EDMONTOSAURUS

0 m   10 m

A herd of *Edmontosaurus* on the move. There is evidence that many dinosaurs such as these lived in large herds and also that they migrated long distances, perhaps thousands of kilometres, in search of the rich food supplies they needed.

Moving in herds

As we scan our eyes across the Late Cretaceous landscape of what is now Alberta, Canada, we see a herd of fifty *Edmontosaurus*, mainly feeding on low bushes, though the larger ones rear up to strip leaves from conifer branches. They chomp constantly, like huge cows, making rasping noises as their jaws pump meditatively and they rearrange the masses of plant material inside their mouths with muscular tongues. Occasionally, an *Edmontosaurus* pauses, raises its snout and gulps down the bolus, the product of all this chewing. Indiscreetly, an *Edmontosaurus* by the side of the river farts, and another drops a huge mass of dung that lands on the ground and flows laterally, forming a hadrosaur pat 2 metres (6½ feet) across. This contains fragments of the dinosaur's last meal, and quickly attracts buzzing flies and some stalwart dung beetles ready to begin the clearing-up operation.

Suddenly, the tyrannosaur *Daspletosaurus* appears on the other side of the river; the pair of *Edmontosaurus* that spot him first rear up on their hind legs, raise their heads and howl. Their cry is taken up by the herd, the larger males uttering deep bellows, the females slightly higher sounds, and the juveniles higher-pitched trumpeting noises. There's safety in numbers on these dangerous plains.

The sheep of the Mesozoic

Perhaps you're thinking here, at last, is a 'real' dinosaur – scaly, scary, colossal. All of the dinosaurs we have met so far were small flesh-eaters, closely related to birds, and so they shared bird-like characters such as feathers and some kind of flight. *Edmontosaurus*, clearly, was not a flyer. It is a hadrosaur, or so-called 'duck-billed dinosaur', a name long applied to these beasts because of the unique shape of their heads, and in particular their long snouts, which broaden at the end, resembling a duck's bill. This broad beak covering was made from keratin, and marked by a series of vertical grooves.

While it might look monstrous to us, *Edmontosaurus* was typical enough for its time – in fact, hadrosaurs were so common in the Late Cretaceous of North America that they are sometimes referred to as the 'sheep of the Mesozoic'. Like sheep, the hadrosaurs seem to have roamed together in large flocks, and like sheep they don't have teeth at the front of their long jaws. Instead, there are two bony plates, one above and one below, for snatching up plant food and passing it further back in the mouth. There, the long jaws are lined with hundreds or even thousands of teeth, for thoroughly shearing the plant matter; hadrosaurs are renowned for this excessive toothiness. They had so many teeth because they ate tough food. In the 1920s, palaeontologists found *Edmontosaurus* specimens apparently with conifer needles in their stomachs. Such tough food also demands a regular supply of new teeth. We have only two sets of teeth in our lifetimes – the baby teeth and the adult teeth, and that's that. Reptiles and fishes, on the other hand, tend to have continuously growing teeth, so when a tooth is worn out it is pushed out of the jaw and a new one pops up from below. In hadrosaurs, this translated into sometimes six or more teeth growing up the inside of the jaw behind each functioning tooth, and the array of teeth formed a remarkable multiple-pointed rasping system, where alternating panels of dentine and enamel in the teeth formed the most complex shearing and grinding armoury seen outside mammals.

This gentle giant grew to between 9 and 12 metres (30 and 39 feet) long – somewhat larger than a sheep – but in the Cretaceous this was by no means a truly large dinosaur, rather falling somewhere near the middle of the size range. *Edmontosaurus* has short, slender arms and massive hind legs with great three-toed feet; these made it a fast runner on two legs, but the hands and arms, too, were adapted for locomotion, with blunt, hoof-like claws.

Opposite:
An *Edmontosaurus* skull, showing the remarkable 'duck bill' at the front, a toothless area used for grasping and tearing plant food, and the massed banks of teeth further back used for chopping the pieces. The huge nostrils, just behind the duckbill, were probably covered by loose skin when the animal was alive, so it could puff and snort and make whistling and tooting noises.

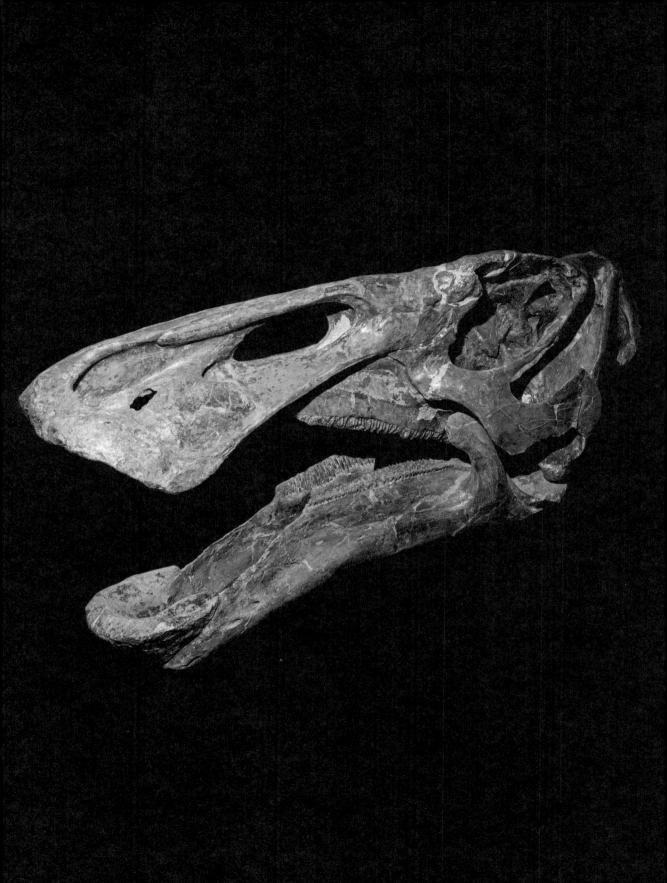

Edmontosaurus had a tough keratin beak over the front of its mouth instead of teeth. This provided a wonderful snipping device, as sharp as scissors, to cut off fronds of conifers or ferns, typical parts of the dinosaur's diet. The ridging on the snout is of uncertain function.

Hadrosaurs are famous for their huge numbers of teeth – as many as 500 in each jaw element, making 2,000 in all. The huge numbers arise because there are many teeth in the functioning row at the top – about fifty here – plus five or six new teeth queuing up below the functioning tooth. They wore through their teeth at a phenomenal rate in chopping tough twigs and leaves, but could replace teeth on a weekly basis.

The ancestors of the hadrosaurs were smaller and truly bipedal, but hadrosaurs like *Edmontosaurus* were so huge, weighing perhaps 4 tonnes (about the same as an adult elephant), that they had to support their weight on their arms when walking or feeding close to the ground. Its tail was massive, and housed the major muscles powering the hind legs; when running, the great caudo-femoral muscle, which attached to the back of the femur (thigh bone) and ran deep down the tail, would contract and pull the leg back.

Amazingly, we also know a lot about the skin of *Edmontosaurus*. It was covered with tiny scales of two types: small, pebbly scales about 1–3 millimetres across, and polygonal scales just under 5 millimetres (¼ inch) across. The small, polygonal scales made up clusters about 2–5 centimetres (1–2 inches) in diameter, and these clusters formed longitudinal rows on the throat, chest and abdomen. The skin with small pebbly or polygonal scales wrinkles around the leg and arm joints. Over the back, the clusters become larger, from 5–10 centimetres (2–4 inches) across. The forearms bear medium-sized polygonal scales, each 1 centimetre (½ inch) across. Down the middle of the back is a crest of rectangular or triangular-shaped scales measuring 5 centimetres (2 inches) tall and forming a regular row, like the crenellated battlements of a medieval castle.

But no feathers.

Dinosaur mummies

How is it possible that we can reconstruct the skin of this long-extinct beast in such intricate detail? In fact, we have had the evidence for over two centuries, in the form of dinosaur 'mummies'. These mummies are skeletons within a body cavity that is composed of sandstone and mudstone but surrounded by a skin impression. In very rare examples the internal structures of individual scales have been preserved. This made the fossils difficult to prepare for

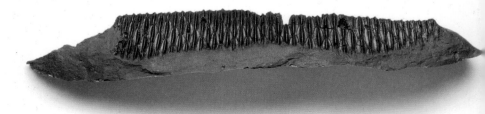

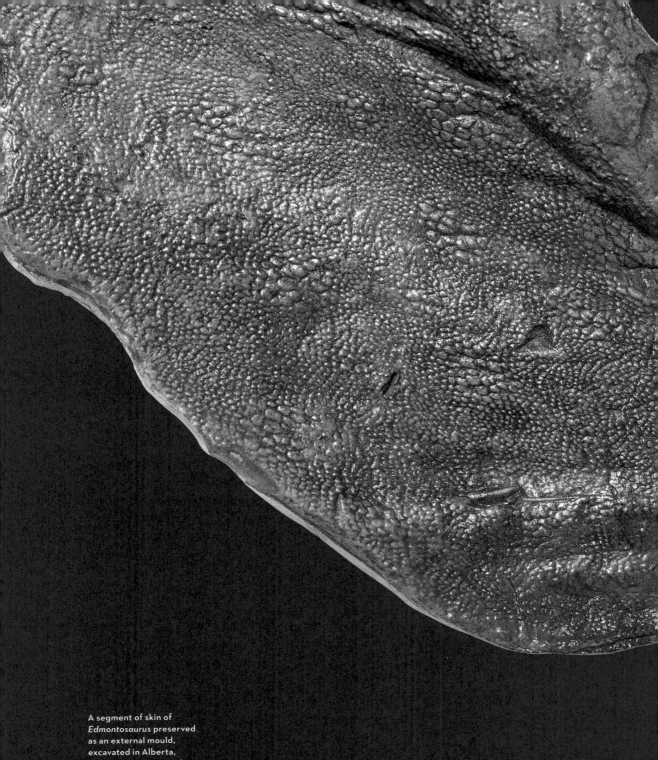

A segment of skin of *Edmontosaurus* preserved as an external mould, excavated in Alberta, Canada. There are scales of varying sizes, clusters of slightly larger scales in patches, with smaller ones between.

museum display, and in the case of some early examples found in the 1890s and early 1900s, much of the body and skin were removed in the lab as preparators dug out the bones.

The first *Edmontosaurus* mummies were excavated in 1908 and 1910 in Wyoming, from the Hell Creek Formation, which stretches across four US states. Indeed the 1908 specimen, part of the collections of the American Museum of Natural History, is still one of the best, preserving skin impressions from all parts of the body except the hindlimbs and tail, which had been eroded away before collection. Further such mummies, as well as integumentary fragments, have been found in the years since, and one that caused a particular media sensation, called MRF-03, was dug up in 2007.

This specimen was found in the Hell Creek Formation of North Dakota, and was described by Phil Manning of the University of Manchester and Tyler Lyson of the Denver Museum of Nature and Science. This work broke new ground. Older descriptions had identified the hadrosaur skin impressions as just that – moulds of the skin surface, not containing any remnants of the skin itself. In his first paper, however, Phil Manning was able to show that in places the skin cells were replicated by inorganic mineral precipitates, and that organic traces and even some structure in the skin survived.

How could these rare organic remains have survived in such a location? This is not a lake with volcanic ash falling or a gentle lagoon, the sites where feathered dinosaurs have been found. The hadrosaur mummies are in fact found in association with many other dinosaur fossils that are simply skeletons, bearing no trace of the skin or any other soft tissues. The location of the 2007 specimen suggests the carcass lay beside a river channel after the animal had died. It might have simply died where it was found, or the carcass might have been washed downriver by a flood and dumped on a sand bar. What followed was crucial. Normally such a carcass would be scavenged by crocodiles, lizards and other carnivores, who would tear off the flesh and skin. Insects would feast on any remnants of flesh left on the bones, and after a week or so nothing would be left but the bones. These, too, would break down under the action of the changing day and night temperatures and rainfall, together with some bone-crushing scavengers, and eventually a mush of bone splinters might disappear into the soil and never be found.

Evidently the North Dakota *Edmontosaurus* specimen, MRF-03, was buried rapidly. It seems it landed on waterlogged soil and sank in. The groundwaters were rich in calcium, reduced iron and manganese ions, and so, as the flesh

One of the classic 'hadrosaur mummies' from Alberta, now in the American Museum of Natural History, New York. This famous specimen was excavated in 1884. Its discoverer, Jacob Wortman, reported that the skeleton was surrounded by a natural cast of the skin, which was unfortunately mostly lost as the workmen struggled to excavate it. The usual approach by bone hunters was to secure the skeleton, and surrounding impressions in the rock were missed or ignored. Nonetheless, there are three patches of moulded skin in the tail region of this specimen.

rotted, carbonate ions replaced the cells of the marinaded skin and muscles. The skin is preserved to a depth of 3.5 millimetres (⅛ inch) and the individual cells fossilized by this mineral cocktail can be seen; there is chemical evidence, too, for the survival of some organic compounds.

Horny dinosaurs and dinosaur dandruff

The main revelation arising from Manning and Lyson's study, and confirmed by a 2019 study by Mauricio Barbi, Phil Bell and others of a similar *Edmontosaurus* specimen from Alberta, is that hadrosaur scales were made of keratinized skin, and had no osteoderms. Keratinized skin refers to the layers that are exposed to the air and are toughened by the protein keratin, which we encountered before (see pp. 48–49) as the component of feathers and hair. This outer skin layer is called the stratum corneum, meaning the 'horny layer', or – to avoid any misinterpretation – the 'keratinized layer'.

There is the same confusion of meanings in Latin, where cor, the noun, means a horn of an animal or a musical horn (think of the cor anglais, or French horn) and *corneum*, the adjective, means 'horn-like' or 'horny', referring to the keratin of a cow's horn, or to a penis. In Ancient Greek, the equivalent word *keras* – the root word for keratin – can similarly mean a cow's horn or a penis. Metaphors never change, and it's interesting that refined agriculturalists have struggled for five thousand years in many languages to describe a cow with large horns…

Skin protects the body, and it grows from inside. When you accidentally graze your skin, the outer, surface layers do not bleed; this is because skin is constantly growing and replacing, and the outer surface is keratinized for protection. As it wears out, flakes fall off, as a constant rain of dandruff.

But reptiles don't get dandruff.

Dandruff forms if your skin is penetrated by feather shafts or hairs. In order for the stratum corneum to be shed, it has to fragment into flakes that are equivalent in size to the average spacing between feather or hair shafts. In a 2018 paper led by Maria McNamara of the University of Cork, we published the first example of dinosaur dandruff, and pointed out this salient difference between endotherms and ectotherms. Endotherms (warm-blooded animals), like birds and mammals, get dandruff, whereas ectotherms (cold-blooded animals), like fishes and reptiles, do not; instead they shed the stratum corneum in great sheets. In fact, it's well known to any herpetologist that lizards and snakes can shed their entire skin (though not really their entire skin, of course – just the thin, keratinized outer layer).

That 2018 paper attracted plenty of attention. We even featured on page 3 of the *Sun* newspaper, the biggest-selling newspaper in the United Kingdom, along with an illustration of a harassed-looking *T. rex* clutching a bottle of Head & Shoulders anti-dandruff shampoo. Of course, they missed that *T. rex* had such silly, short little arms it could not have reached its head.

Trumpeting dinosaurs

Hadrosaurs honked. Palaeontologists don't know this for sure, but it's widely agreed. The evidence comes from the fact that all hadrosaurs had expanded nasal cavities. In some cases, these extended out over the roof of the snout, or even back over the cranium, as an array of crests, some shaped like upended plates, others like snorkels.

In the 1980s, David Weishampel and colleagues made a model of the skull of one of the more extreme crested forms, *Parasaurolophus*, which lived at the same time as *Edmontosaurus*. They modelled the interior nasal skin and then blew into their contraption. In life, the hadrosaur breathes in and air goes through the nostrils at the tip of the snout, through the nasal cavity and down the throat. Breathing out reverses this. When there's a crest, the air goes up through winding passages through the crest, and the length and pattern of how the nasal passages wind around defines the sound. Weishampel reported that Parasaurolophus made a sound like a crumhorn (or *krumhorn*), a medieval wind instrument known especially from Germany that made a strong buzzing sound.

The hadrosaurs variously tooted and parped, and it seems in some cases the tone of the toot varied with body size, so babies might have trilled, adults parped, and males and females may even have had different tones. In the Late Cretaceous, another five or six hadrosaur species would have shared the Alberta landscape with our herd of *Edmontosaurus*. As they move about they each toot and trumpet with their own timbre, the *Edmontosaurus* making their low sounds by inflating the skin behind their nostrils.

EOMAIA

0 cm   10 cm

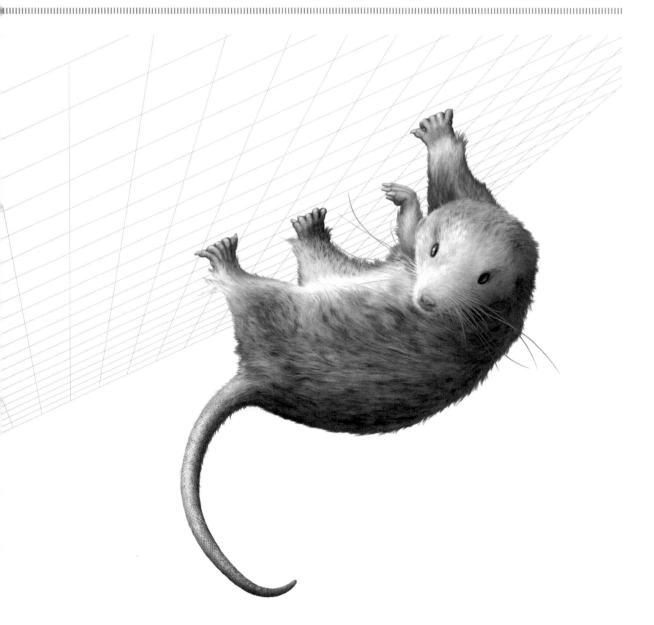

The first mother

We have already visited the Early Cretaceous in north-east China several times, to see *Sinosauropteryx* (pp. 28–41), *Caudipteryx* (pp. 56–69), *Microraptor* (pp. 98–113) and *Confuciusornis* (pp. 70–83). As these dinosaurs and early birds ran up and down the tree trunks, and glided and fluttered from spot to spot, other animals watched them attentively. They did well to keep an eye on these predators that could swoop and kill in an instant.

On the ground, two little mammals jump and squeak. They seem to be playing, but who can tell for sure? They rush past each other, jumping up onto a dead branch, and swinging under, clutching the rough surface of the bark with the sharp claws on their pink little hands.

These are *Eomaia*, distant ancestors of all the mammals alive today. The long limbs and fingers show that these little mammals were climbers, just like squirrels, with powerful arms and legs for gripping, and long claws to dig into the bark. They likely charged around in the trees in pursuit of insects.

Only a single *Eomaia* fossil has ever been found, but it is amazingly complete: it not only shows all these details of limbs and jaws that help us reconstruct their mode of life, but it is also a key fossil for understanding mammal evolution. The name *Eomaia* means 'first mother' and refers to the fact that these were the oldest placental mammals; we are placental mammals, so this tells us something about our distant ancestry. But first, the fossil.

The fossil

The *Eomaia* fossil was first published in 2002. The animal is just 10 centimetres (4 inches) long and weighed an estimated 20 to 25 grams (0.7–0.9 ounces), about the size of a gerbil or mouse. The head points down, and the left arm and leg are stretched out, showing every bone, including the delicate fingers and toes.

The backbone and rib cage are clearly seen, as is the base of the tail. The tail likely would have been about twice the length seen in the fossil, but a thin layer of sediment covers up its end. In fact, you can see the edge where the technician has chipped off this thin rock layer to expose the hand and foot, and carefully removed rock flakes round the back to expose the fur. The mushy, dark layer under the rib cage that extends over the spine is

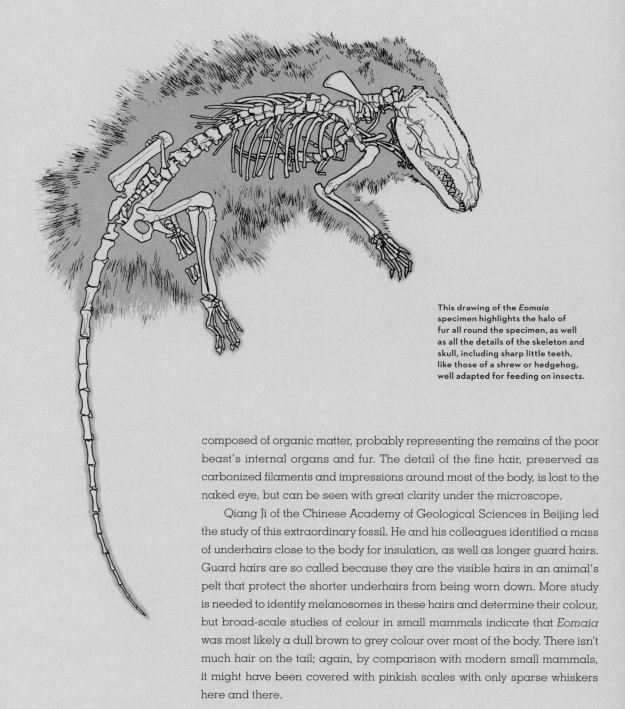

This drawing of the *Eomaia* specimen highlights the halo of fur all round the specimen, as well as all the details of the skeleton and skull, including sharp little teeth, like those of a shrew or hedgehog, well adapted for feeding on insects.

composed of organic matter, probably representing the remains of the poor beast's internal organs and fur. The detail of the fine hair, preserved as carbonized filaments and impressions around most of the body, is lost to the naked eye, but can be seen with great clarity under the microscope.

Qiang Ji of the Chinese Academy of Geological Sciences in Beijing led the study of this extraordinary fossil. He and his colleagues identified a mass of underhairs close to the body for insulation, as well as longer guard hairs. Guard hairs are so called because they are the visible hairs in an animal's pelt that protect the shorter underhairs from being worn down. More study is needed to identify melanosomes in these hairs and determine their colour, but broad-scale studies of colour in small mammals indicate that *Eomaia* was most likely a dull brown to grey colour over most of the body. There isn't much hair on the tail; again, by comparison with modern small mammals, it might have been covered with pinkish scales with only sparse whiskers here and there.

But how could *Eomaia* be identified confidently as a placental mammal? The fossil cannot preserve evidence of the womb; in fact, it's impossible to be sure even whether the specimen is a male or female. To answer this question, we need to look at the differences between placental mammals and marsupials.

Mammal reproduction

By 1800, biologists recognized that there were two main groups of mammals: placentals and marsupials. This realization shocked the European establishment. Marsupials had been seen by Europeans mainly in Australia, and we can only imagine the consternation of these eighteenth-century explorers when they first saw a kangaroo. In 1699, William Dampier wrote of the smaller wallabies: 'The Land-Animals that we saw here were only a sort of Racoons... for these have very short Fore-Legs; but go jumping upon them as the others do (and like them are good meat).' Note that nearly all the early reports of marsupials were written more in a culinary than a zoological context.

Biologists noted the pouch, and the tiny baby marsupials that occupied it. It is this pouch that distinguishes the marsupials on the one hand from the placentals on the other (and gives the marsupials their name, from the ancient Greek *mársippos*, meaning 'pouch'). Marsupials give birth to undeveloped embryos that crawl, blind, but with proportionally huge arms, up from the mother's cervix into the pouch, and then develop by feeding on milk through teats in the pouch. Placental mammals, such as humans, cats and cows, allow their young to develop to a much more advanced stage in the womb, feeding through the placenta, hence 'placental mammals'.

It is worth noting that there is a third group of living mammals. When the first dead duck-billed platypus was sent from Australia to Europe in 1799, English zoologist George Shaw remarked, 'It naturally excites the idea of some deceptive preparation by artificial means' – in other words, sticking the beak of a duck on the body of a beaver or giant mole. But its unusual bill is not the important point; echidnas, relatives of the platypus, look more like pointy-nosed porcupines. It's the fact that they lay eggs, presumably a primitive characteristic inherited from reptilian ancestors. The platypus and the echidnas are living monotremes, meaning 'one hole', so called because the urinary, defecatory and reproductive tracts are united through a single opening, the cloaca.

Following pages:
When the first duck-billed platypus skins and skeletons reached Europe they were treated as monstrous fabrications. This is the first published illustration, dating from 1799, based on drawings by Frederick Nodder in George Shaw's *The Naturalist's Miscellany, or Coloured Figures of Natural Objects*, Volume 10. This provided evidence that a very few mammals still laid eggs, a shocking concept to naturalists who were familiar only with the mammals of the northern hemisphere.

385. Pub,d by F P Nodder June 1799,

The oldest mammals

So, having identified these three groups of living mammals, and having worked out that the egg-laying monotremes were likely the most primitive, the hunt was on for fossils. The first examples came as a surprise to William Buckland (1784–1856), who was Dean of Christ Church and Professor of Geology at the University of Oxford. By 1820, Buckland had already built a reputation as a great field naturalist as well as an eccentric. He had investigated bone caves in Yorkshire and southern England and early human bones, which he understood by reference to the Bible. He was also known for his practical jokes, including using bones as blackboard pointers while teaching and offering his guests roasted moles or tiger steaks when they came to dinner.

In 1824, Buckland named the first dinosaur ever to be reported, *Megalosaurus* from the Middle Jurassic, discovered close to Oxford. Associated with the huge fossils of this beast were some tiny jaws, which at that time were thought to have come from marsupials. Both finds were amazing, but in some ways the tiny jaws, later identified as the mammal *Amphitherium*, were the most startling.

In the nineteenth century, geologists divided up the history of life into three major stages, termed the Primary, Secondary and Tertiary. The Primary rocks contained the most ancient life forms, which looked very little like anything now living and, as for vertebrates, it was the 'age of fishes'. The Secondary – what we now call the Mesozoic – was the 'age of reptiles', named after the marine reptiles found in different places and the bones of large crocodiles and other unknown reptiles. The Tertiary was the 'age of mammals', attested in excellent fossils from the Paris basin and Germany. So, finding a mammal among Mesozoic remains broke the general picture of a progression of types through time, from fish to reptile to mammal. Of course, Buckland, as a good Christian, viewed this succession of forms not as evidence for evolution, but for a series of floods, of which the last was Noah's Deluge.

Opposite:
An example of one of the first mammals on Earth, *Megazostrodon* from the Early Jurassic, some 190 million years old, of South Africa, showing the trunk region with the right hind limb and foot viewed from below. *Megazostrodon* is thought to have been a mouse-sized, insect-eating, nocturnal mammal, but still almost certainly an egg-layer.

Over the next century, mammal fossils were reported – albeit infrequently – from the Jurassic and Cretaceous of other locations in Europe, and then North America. By 1929, when the great twentieth-century palaeontologist George Gaylord Simpson (1902–1984) reviewed everything that was known about these early mammals, looking at the tiny, priceless specimens from many museums around the world, he commented that they would all quite easily fit in his hat. Admittedly, Simpson did have a large head, and so presumably rather a large hat, but that does not detract from the fact that fossils of Mesozoic mammals were sparse.

Not much changed during the rest of the twentieth century, although new finds were identified from new parts of the world, such as South America and Australia, and a few possible mammal fossils were even reported from the Late Triassic. It seemed that, indeed, the modern mammal groups were primarily post-Mesozoic in age. The textbook position was that while various primitive types of mammals had existed in the Jurassic and Cretaceous, they couldn't be classified as any of the three modern groups; rather, these had evolved in a huge diversification event after the demise of the dinosaurs.

When the dinosaurs disappeared, killed off by a huge asteroid impact 66 million years ago, the mammals took over, and they did so rapidly. All the modern monotremes, marsupials and placentals emerged at that time, to occupy the ecological space vacated by the dinosaurs.

Debates and squabbles

This basic chronological orthodoxy was challenged head on in the 1980s. While palaeontologists bemoaned the sparseness of mammal fossils through the Mesozoic, a new breed of scientists decided they could largely do without them. These were the molecular biologists, who realized that they could sequence the proteins and DNA of plants and animals and discover their relationships according to the amount of accumulated difference between, or shared mutations in, the genetic sequences. In so doing, they could trace the distant origins of species.

The first estimates from these new studies placed the origin of the major groups of mammals some time in the Late Jurassic and Early Cretaceous – about twice as early as the ages of the oldest fossils. And those projected ancient origins were for early members of modern groups such as monkeys, cows and cats. I remember being incensed at such an absurd suggestion – cats, cows and monkeys living side by side with dinosaurs like *Stegosaurus* and *Psittacosaurus*! Both sides exchanged heated essays and commentaries, and squabbled at scientific meetings: the fossil evidence must be right, claimed the palaeontologists; the molecules never lie, claimed the gene sequencers.

As it turns out, the molecular dates were partly right and partly wrong, and the palaeontologists were also partly right and partly wrong. The current consensus is that indeed there was an explosion of modern mammal groups immediately after the extinction of the dinosaurs 66 million years ago, and that's when the first monkeys, cows and cats appeared. However, there was also a long, unexpected history of primitive placentals and marsupials right back to the Jurassic, for which the fossils had been nearly entirely absent, apart from odd teeth and jaws from the Late Cretaceous that had proved highly controversial.

The discovery of fossil mammals from the Late Jurassic and Early Cretaceous of China caused as much of a shock to the scientific world as the first reports of feathered dinosaurs, starting with *Sinosauropteryx* (see pp. 28–41). Three fossil mammal specimens were reported between 1997 and 2000, and these were astonishing, preserving the whole skeleton – not just isolated little jaws, as George Simpson with his hat had contended with. These mammal fossils from China belonged to extinct mammal groups, and not to placentals, marsupials or monotremes. They were beautiful, they were complete, and they provided untold new information – even showing traces of fur – but they did not disturb the assumption that the modern groups were probably not Mesozoic in age.

That was until the reports of *Eomaia*.

Eomaia possesses a great secret. This tiny furball would have been nearly invisible if we travelled back in time to the Cretaceous. It probably hunted at night, and would have relied on its silent movements to escape predation by the abundant tree-dwelling birds and theropod dinosaurs of its day. But, although still an egg-layer, this was one of the first placental mammals, founder of our mammalian dynasty, the 'first mother', as her name says.

Where *Eomaia* fits

Although classified as a placental mammal, *Eomaia* might have had a more primitive reproductive system. Its hip bones are narrow, and it has a pair of tiny bones that project from the front of the pelvis called epipubic bones. In marsupials, the epipubic bones help to support the pouch, so could *Eomaia* have had a pouch and given birth to undeveloped embryos? Very likely.

This might seem a killer argument that *Eomaia* is not a placental mammal, but there are other clues, especially certain features of the ankle and jaws that are distinctive to modern placental mammals. It is more primitive, however, than all living placentals, which is no surprise, lacking for example further specializations of the ankle, as well as having a primitive tooth formula. Modern placental mammals have three or fewer incisor teeth on each side (humans have two), but *Eomaia* has five upper and four lower incisors. In other words, it retains quite a few primitive features seen also in early marsupials and even in the Mesozoic mammals that don't belong to any of the modern groups.

The conundrum is resolved in two ways. In fact, all of the early mammals seem to have either laid eggs like modern monotremes or had pouches (based on the evidence of the epipubic bones). Retention of such features of ancestral mammals in an early placental is fine because evolution doesn't happen instantly. Just as the first birds retained teeth and the long bony tail of the ancestral dinosaurs, so the first placental mammals might well still have laid eggs or given birth to tiny babies that developed further in a pouch. The second point is about semantics – perhaps we shouldn't call the placental mammals 'placental mammals', since they didn't all have a placenta. But that was the name they were given ages ago, and we can't stop people using a term that is true for all the living forms at least. The fossils show us that the placenta might have arisen later, maybe in the Late Cretaceous, while the early placentals retained the breeding systems of their ancestors.

Since *Eomaia* was announced in 2002, an even older placental mammal, *Juramaia*, has been reported, also from China, from the Middle to Late Jurassic, so perhaps 164 million years ago. Its preservation is poorer, and evidence of hair is sparse, but the fossil confirms that the three modern mammal groups had independent histories of at least 100 million years. These small, climbing animals clung onto survival, scuttling through the trees and feeding on insects, until they had their chance 66 million years ago, with the disappearance of the dinosaurs, to diversify into all the familiar modern groups.

9

SALTASAURUS

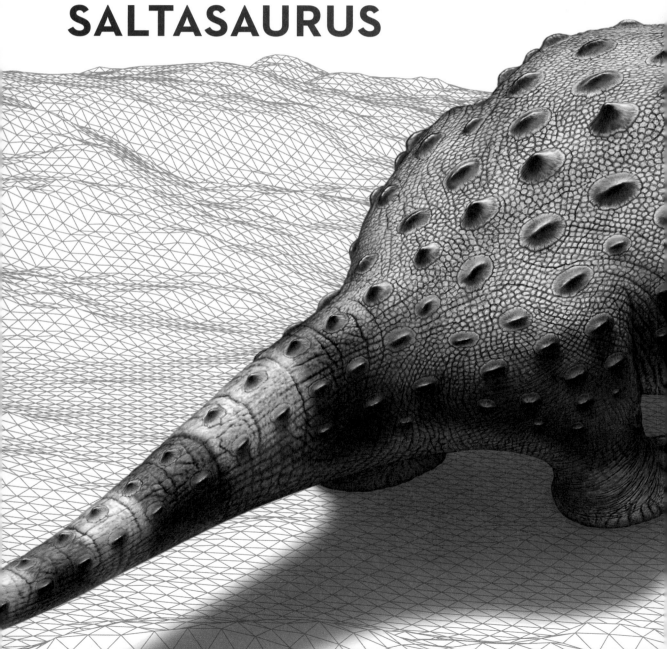

0 m   13 m

72–66

Cross-section through a monkey puzzle cone, *Araucaria araucana*. This fossil comes from the Jurassic of Argentina, and it shows that the cone is composed of hundreds of seeds that can be spread over a wide area after the cone falls and is perhaps eaten and dispersed.

Opposite:
Modern *Araucaria* forests in Chile and southern Brazil replicate the scene in South America about 70 million years earlier. Species of the monkey puzzle occur also in Australia, New Guinea and New Caledonia, evidence for long-term survival of this tree over the southern hemisphere.

Armoured monsters

We have travelled back in time some 70 million years to what is now Argentina, which was located slightly further north in the Late Cretaceous. Mountains rise above the habitable plains, and fast-flowing rivers transport sand from their higher reaches. In the arid summer, dune fields spread out across the landscape and dinosaurs retreat to the wetter areas around the rivers and pools, where there is low scrub, and conifers, such as monkey puzzles, are abundant, as well as ferns and seed ferns. The new flowering plants are there too, magnolias, roses and vines.

A gang of the leggy, flesh-eating *Noasaurus* scamper past, and a small flock of turkey-sized birds, *Enantiornis*, settle on the ground and peck at seeds. They peer about as they peck, and then as one they all raise their heads up, startled, but can see nothing. The leaves rattle on the bushes. The dust rises and billows. Something is coming.

On massive feet, a mother and daughter *Saltasaurus* appear from behind a hillock, and the birds scatter. But they need have no fear. *Saltasaurus* are plant-eaters, and they are constantly on the move because they have to find 100 kilograms (220 pounds) of plant food each day in this dry, sparsely vegetated landscape.

The adult *Saltasaurus* is 12.8 metres (42 feet) long and weighs about 7 tonnes. Her daughter is a third of her length, but weighs only about one-tenth as much as her mother. They snatch at whole bushes and trees with their long jaws, stripping branches bare; the teeth on the upper and lower jaws of their long snouts point forwards at a 45-degree angle, enabling them to grasp a bunch of leaves, as we grasp things with our thumb and forefingers, and keep the jaws together while they pull their mighty heads backwards. As they step away from the tree, they neatly strip the leaves off the branch, using the teeth as rakes. Gulp, down goes a great mass of leaves, entirely unchewed, and they move on.

An *Enantiornis* perched high in a tree watches the pair of *Saltasaurus* from above. Their leathery, studded backs are of particular interest. Large, bony studs are set in neat rows from front to back, and between these are numerous smaller interlocking plates. The *Enantiornis* flops down from the tree, and lands clumsily on the mother's back.

Three views of a *Saltasaurus* osteoderm (bony plate set into the skin). In external view (1), the central boss and radiating sculpture can be seen. The edges mark out a six-sided figure, caused by the fact that the osteoderms fit together snugly against a series of tiny bony plates round the edges. The central boss is clearly seen in the side views of the osteoderm (2, 3), evidence of the armour-like function. The bony chainmail covering the entire body would be difficult for a predator to penetrate; it would likely break its teeth or claws on these bosses.

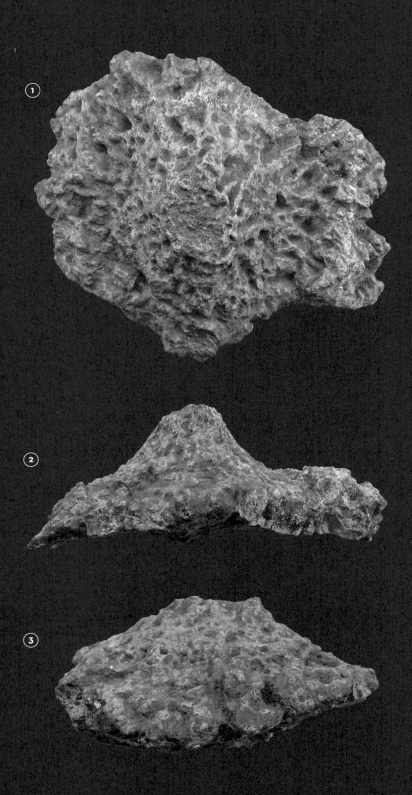

The *Saltasaurus* mother plods on, entirely unaware that she has acquired a passenger. The *Enantiornis* pecks around, looking for seeds and insects that have slipped between the plates. It plucks a juicy, blood-filled louse from the soft skin between the small armour plates, and the *Saltasaurus* gives a grunt of pain, or perhaps relief. She is unable to keep herself free of parasites and has learned to live with the constant, distant pain. Her only defence is to roll in the sand, or mud, but that is a large effort, and there is a risk she might get stuck in a pool and be unable to right herself or scrabble back up the crumbling bank of a river. Her daughter has no such qualms and is forever rolling and scrambling in and out of ponds and mud pits.

The first giant dinosaur with armour

There is no complete *Saltasaurus* skeleton. The original specimen that gave the species its name, described in 1980 by two renowned Argentinian palaeontologists, José Bonaparte and Jaime Powell, was a fossil of the hip region that had been excavated from the Lecho Formation of north-western Argentina. But they also found many other isolated bones, and armour plates. Indeed, Bonaparte and Powell chose as the species name *loricatus*, which means 'protected by small armour plates'.

However, by 1992, after more collecting, Powell could identify five *Saltasaurus* individuals among the remains, which included several skulls and bones from most parts of the skeleton. Still, none was complete, and so Powell studied other titanosaurs, giant sauropods (large plant-eaters) that occupied most of the world, but especially South America, Africa, India and China in the Late Cretaceous, to gain a detailed understanding of the whole animal.

As sauropods go, *Saltasaurus* was small, with a short neck and tail, but extremely broad in the beam. The original fossil of the hip region showed that, and this is a characteristic shared by all titanosaurs – the torso is wider than tall, and the arms and legs are spaced very wide apart. In fact, most other sauropods were narrow-hipped, and left prints relatively close to the midline of their direction of movement when they walked. But titanosaurs plonked their feet down metres apart, leaving what seems like two separate successions of prints, one for the left feet and one for the right.

As for its diet, fossilized dung of a titanosaur from India showed fragments and pollen grains from a broad range of plants, including cycads and conifers, as might have been expected. But the pollen showed that the Indian titanosaur had also eaten palms and grasses, so it was feeding on some of the new flowering plants, which had emerged in the mid-Cretaceous. Although they now dominate the world, these plants were much less abundant back in the Cretaceous, and other evidence had suggested that they did not form a significant part of the diet of dinosaurs. More dung is required.

The armour plates were wholly unexpected. Powell noted that the discovery of osteoderms in 1980 was a first for sauropods. However, they have since been found in titanosaurs from other parts of the world. Powell distinguished bony armour plates and so-called 'intradermal ossicles' in the fossils. The bony armour plates are about 12 centimetres (4¾ inches) across – roughly the size of a human hand – and they rose to a point in the midline or a rounded ridge to the sides. The ossicles, by contrast, are only 7 millimetres (¼ inch) across, smaller than the nail on your fifth finger, and they are roughly circular in shape. The ossicles were packed into the outer layers of the skin, each separated from its neighbours by a narrow bit of tough, but flexible, skin.

The larger armour plates would have deterred predatory attacks, and the smaller ossicles between them formed a perfect chain-mail armament, impossible for any dinosaur to bite through – but vulnerable to pesky parasites. Probably, just as modern whales have to withstand all kinds of annoying skin parasites, such as sea lice, as well as barnacles that treat the whale as a handy floating island, sauropod dinosaurs contended with lice, fleas and gadflies that fed on their blood. So long as these irritating beasts kept out of the way, the dinosaur could not reach them, and would have relied on birds or pterosaurs to snatch the nourishing, blood-filled arthropods off their host.

Saltasaurus chainmail: the small bone plates are nearly 1 centimetre (½ inch) thick, and they fit snugly around the larger bosses, which are tipped with a bony spine covered with a keratin-horn sheath. Large predators would retire wounded if they tried to penetrate this armour, whereas probably some smaller, even parasitic, organisms might have been able to squeeze through the cracks.

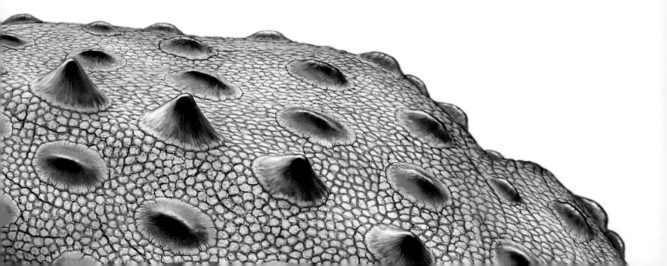

Microscopic cross-section of a smaller bony plate showing a plywood-like construction of vertical and horizontal bundles of fibres, shown in orange and blue in this enhanced, fluorescent image. The crossing fibre bundles mean that this bone structure is almost impossible to break, whichever direction you try to attack.

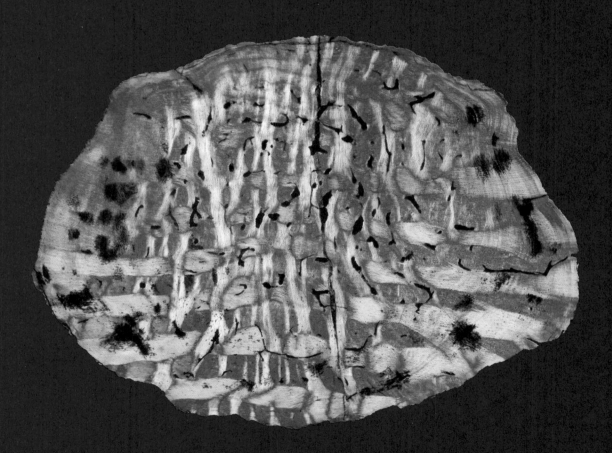

Parental care

No *Saltasaurus* nests or babies are known, but a locality in Argentina called Auca Mahuevo has yielded rich evidence of the parenting practices of other titanosaurian species. The site was appropriately named by its discoverers, Luis Chiappe and Lowell Dingus – 'mas huevos' means 'more eggs' in Spanish, as hungry members of the dinosaur crew called to the camp cook.

And there were more eggs. Chiappe and Dingus identified several hundred nests, each with about twenty-five eggs. They suggested that this had been a favoured nesting place, where the mother titanosaurs returned each year to lay their eggs. They scooped trench-like hollows in the sand, dropped in the eggs and then scuffed earth and sand back over them, to offer some kind of protection.

The eggs are spherical and small, only 13–15 centimetres (5–6 inches) across – about the size of a large goose egg – so the titanosaurs that laid them were likely small as well. In most cases, the nests were either empty or contained only broken pieces of eggshell – the babies had evidently hatched and gone.

Most unexpected was that some of the eggs contained the remains of tiny embryos, including pieces of their skin. Even before they hatched, the babies had armour plates in their skin: tiny beads of bone, each as little as one-tenth to half of a millimetre across. These extraordinary traces of embryonic skin showed all kinds of arrangements of the little bony plates – parallel rows, rosettes and flowers.

These patterns, while being rather beautiful, give some clues about the development of the armour plates, suggesting they matched one-for-one the pattern of scales that would otherwise cover the body. This kind of patterning relationship between scales and bony plates is seen in modern crocodilians and lizards, where the patterns may be directly genetically determined. Biologists have discovered in such cases that certain patterns are common, and others are forbidden – they are never seen and cannot be reproduced in any healthy animal. The same is likely true of dinosaurs, and so the plate patterns may be saying something fundamental about their genetic codes – but we have yet to learn how to read it.

Details of the *Saltasaurus* eggshell, under the microscope. The concave underside of the eggshell fragment (1) shows bumps known as mammillae. The fracture edge to the left of the image has numerous pores running through it. These mammillae and pores are connected and function for gas exchange (the embryo has to absorb oxygen and dispose of carbon dioxide) and also partly as a protective device. The external surface of the eggshell (2) shows the roughened, protective texture.

Impossible creatures

Palaeontologists have long been puzzled about how giants such as *Saltasaurus*, and its fellow sauropods, could have survived. Although *Saltasaurus* might only rarely have exceeded 10 tonnes, that is still twice the weight of a bull African elephant; other sauropods could grow up to 30 metres (100 feet) long and weigh 50 tonnes. How could any land animal achieve, and maintain, such vast size?

We could imagine that these dinosaurs inhabited a different world, with different physical properties from our own. Maybe the planet's gravitational pull was weaker, or perhaps these behemoths lived half their lives in water, allowing their huge bulk to float. But there is no evidence for either of these guesses. In fact, the Earth was the same size in the Jurassic, and so its gravitational force must have been the same; and the fossilized trackways of sauropods show that they were walking over the land, not floundering through deep ponds and seas.

The only major difference then was that temperatures worldwide were hot to temperate, with no polar ice caps. This meant that dinosaurs could wander nearly anywhere and find similar climatic conditions, with hot summers and mild winters, and this was key to their survival strategy. The secret to their success was that sauropods and other large dinosaurs had all the advantages of being warm-blooded, as we are, but without the costs.

Mammals and birds expend the vast majority of their energy maintaining a constant body temperature. To do this, they typically eat about ten times as much as a cold-blooded animal, such as a lizard or crocodile, of the

The *Saltasaurus* tail carried armour right to the tip. Why did *Saltasaurus* have a tail? It wasn't used for balance, in the way bipeds wave the tail to prevent themselves from toppling onto their noses. Perhaps, like cows, *Saltasaurus* thrashed it around to deter insects. But half the tail was needed to accommodate the main leg muscles.

same body mass. In other words, nine-tenths of what we eat is burned up by our internal furnaces, maintaining a high and constant body temperature. Only one-tenth of our energy intake is devoted to the maintenance of other bodily functions and movement. So, a sauropod dinosaur, weighing 50 tonnes, could have survived by consuming the same number of calories as a 5-tonne elephant.

The Argentinian nests also provide a clue to the sauropods' evolutionary success. Placental mammals carry their young to an advanced stage of development, and even after they are born, the parents may devote a great deal of effort and time to feeding and educating their offspring. A female elephant is typically pregnant with a single calf for nearly two years, the birth process can be dangerous, and the infant requires several years of constant supervision and rescuing from scrapes such as falling into holes or being attacked by predators before it is able to survive on its own.

Mother sauropods typically laid a clutch of ten or twenty eggs and departed, just as they did at Auca Mahuevo. The eggs were rarely larger than a football or rugby ball, so the effort in producing the eggs was modest, and the young were left more or less to fend for themselves. Even if most of them died, the species would survive at replacement level, and the mother could always lay more eggs. This strategy is another great energy-saver, and the end result is the same – although arguably the dinosaurs were probably more stupid than the elephants, so they learned less. Ignorance is bliss.

Saltasaurus was a giant, as this scaling diagram shows. However, it was by no means the largest of the sauropods – some of them were 30 metres (99 feet) long and perhaps weighed 50 tonnes. Like elephants today, the sauropods were so large they were more or less immune from predation. Except they did need armour, and this might have been to act as a deterrent against attackers who could have made a nuisance of themselves even if they could not have killed the *Saltasaurus* outright.

6 m

5 m

4 m

3 m

2 m

1 m

0 m

10

PSITTACOSAURUS

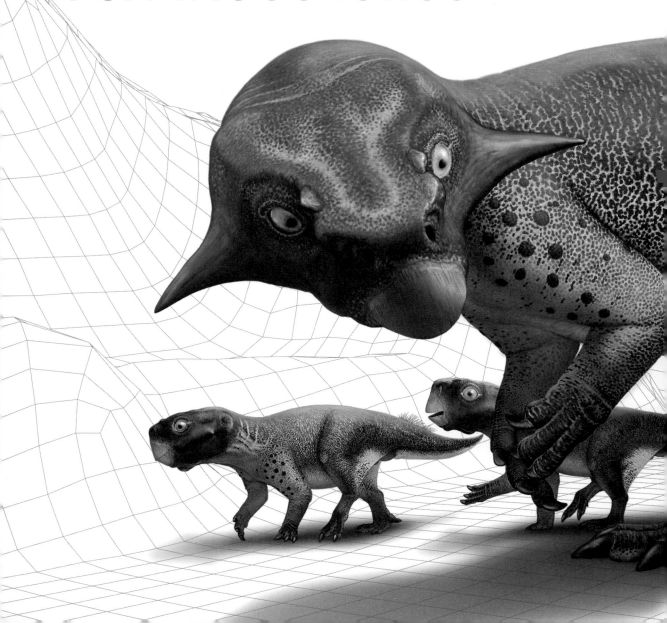

0 m   2 m

130–120

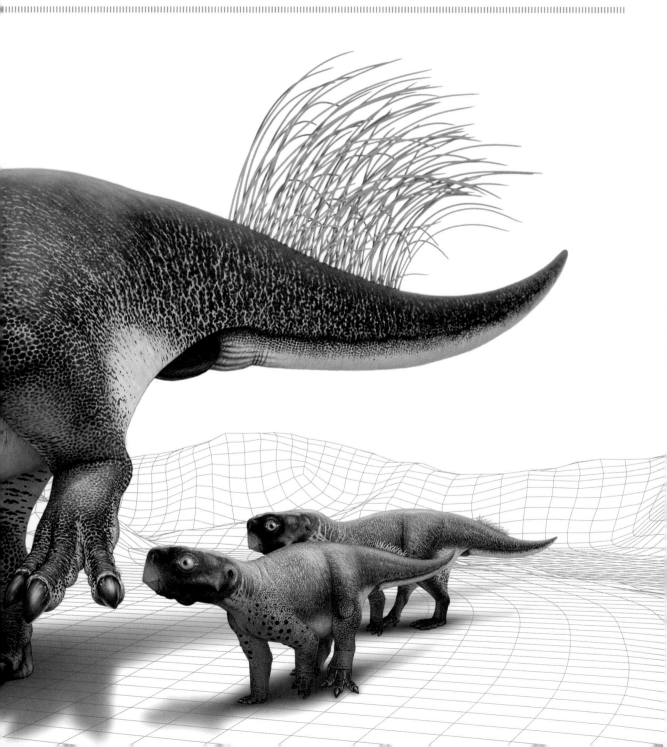

Hiding in plain view

If we went back 125 million years to north China, we might stand in wooded glades, punctuated by wide, shallow lakes, with volcanoes belching smoke in the far distance. From the bright birds and small dinosaurs in the trees to the large insects flying out over the lakes, everything is lively and noisy. Every now and then, something rustles through the trees, but it's hard to see what it is. Moving closer, you find a small herd of snub-nosed dinosaurs, each about 2 metres (6½ feet) long, and seeming to flit mysteriously from spot to spot.

These are *Psittacosaurus*. Babies stay on all fours but the adults walk on their hind legs, and have a heavy head, flattened at the front, where the beak-like snout curves into a sharp tip. They use this short beak to snip leaves from the trees and chop them further back in the mouth. But it's their ability to blend into the background that is most striking. They seem curiously two-dimensional; then the light flickering through the trees changes, and they are gone.

Clever camouflage

This camouflaging effect is created by countershading, which works by balancing the effects of sunlight. If the belly is in shadow under natural sunlight on an open plain, by being pale it matches the darker, but more brightly lit, back, giving the animal a two-dimensional appearance so that it effectively disappears into the background. In closed habitats, such as forests, the back is not so brightly illuminated as the sun is partially blocked by foliage, and the shadows on the underside of the animal provide less contrast, so the flattening effect is produced by making less of the sides pale.

In a game-changing publication in 2016, Jakob Vinther and Innes Cuthill from the University of Bristol showed that the scaly skin of *Psittacosaurus* was countershaded. Cuthill, a biologist, has studied animal coloration all his life, but he is used to working with living animals, carrying out experiments on the function of their colours and patterns. Vinther is a palaeontologist who has figured strongly in many areas of palaeocolour reconstruction (see p. 50). Having seen a *Psittacosaurus* specimen in the Senckenberg Museum in Frankfurt, Germany, Vinther asked Cuthill whether different kinds of countershading in the dinosaurs might indicate something about the environments they lived in. He was convinced the original colours were

The adult *Psittacosaurus* skull shows
the deep jaws lined with chopping teeth.
There are no teeth at the front of the jaws,
so this animal presumably fed on tough
vegetation that it cropped with its bony,
horn-covered beak. It likely had a muscular
tongue to pull the cropped plants back and
then chopped them with its sharp cheek
teeth. The skull is deep and broad at the
back, providing great space for powerful
jaw-closing muscles.

The famous Senckenberg Museum *Psittacosaurus*. This is an unusual specimen because it preserves soft tissues. (Most examples are preserved in volcanic ash and any feathers or soft tissues were burned up.) The black rim round the skeleton is largely the skin of the animal, including the array of reed-like feathers on top of the tail (see the left-hand side of the tail in the fossil). In detail, all the blackened areas show scales and other features of the skin, including hints of the original colour.

preserved in the scales, or at least the discrimination between dark and light colours. The Senckenberg *Psittacosaurus* had a light underbelly and tail and a more pigmented chest, while the skin over the head and back carried dark-coloured scales.

When these colour shades were projected onto an accurate 3D model constructed by Bob Nicholls, they showed a low position for the transition from pale to dark scales along the side of the body and tail. By comparison with modern mammals in their native habitats, this indicates *Psittacosaurus* lived in a mixed, forested habitat, as is suggested by the fossil plants often found alongside it. If the pale belly had extended further up the beast's sides, that would have suggested that the creature lived on a more open plain.

Psittacosaurus was a very successful dinosaur. Since its discovery in 1923, eighteen species of *Psittacosaurus* have been named, coming from dozens of localities scattered over eastern Asia, from southern Siberia through Mongolia, north China, Korea, Japan and Thailand. Perhaps only ten of these eighteen species are currently regarded as distinctive, but nonetheless this shows an unusually rich diversification of a single dinosaur genus over a wide geographic area.

An unexpected discovery

Psittacosaurus was first named in 1923 by Henry Fairfield Osborn (1857–1935), the somewhat imperious Director of the American Museum of Natural History in New York, who had sponsored a series of palaeontological expeditions to Mongolia in search of the origins of humankind. His expeditions did not find any early humans, but they did bring back a rich collection of Cretaceous dinosaur fossils.

Osborn set about writing up the finds, giving names to various dinosaurs. He saw straightaway that *Psittacosaurus* was a most peculiar animal. It was clearly a ceratopsian dinosaur, an early relative of famous forms such as *Triceratops* and *Monoclonius*, great four-footed animals with horns on their noses and over their eyes and a substantial bony frill that sprouted from the back of the skull over the neck. The ceratopsians were familiar and abundant dinosaurs in the Late Cretaceous of North America, but they were all huge and they all walked on four legs, like overgrown rhinoceroses. *Psittacosaurus* was small, and adults were bipedal. Here was a kind of 'missing link', Osborn

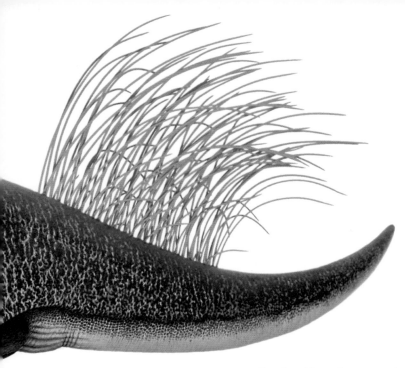

Swamp reeds or feathers? When the vertical bristles along the middle of the *Psittacosaurus* tail were first described, some palaeontologists thought that fossil plants had become associated with the dinosaur skeleton. However, there is no doubt these are part of the animal, and that they are made from keratin, and so really are bristles or feathers. Were they used in some way as a signal from individual to individual?

realized, midway between the small, bipedal, unarmoured dinosaurs of the Jurassic and the frilled and horned ceratopsians that already filled the galleries of his museum.

Then, something very unexpected was reported in 2002: the Senckenberg *Psittacosaurus* was found to have had feathers. This came as a great surprise; after all, this was an armoured dinosaur, a long way from birds in the evolutionary tree. The 'feathers' themselves were equally bizarre, a fence of tough, reed-like structures arranged along the midline of its back.

This specimen, from Sihetun, China, is so remarkably well preserved that great sheets of its scaly skin survive more or less in place. After it died, the animal must have slumped to the bed of a shallow lake, and the flesh rotted or was scavenged, but the skeleton remained undisturbed, with the dorsal bristles in place, and sheets of skin over the bones. The skin has been preserved by crystallization of the calcium phosphate (apatite) that was partly in the skin and from surrounding tissues including the bones. This is a very unusual kind of preservation.

Gerald Mayr and co-authors from the Senckenberg Museum hedged their bets when they presented their findings to the public in 2002. They called the features 'integumentary structures', going on to say that there was 'no convincing evidence which shows these structures to be homologous to the structurally different integumentary filaments of theropod dinosaurs' – in other words, feathers. So far as we know, *Psittacosaurus* did not have any other feathers; only these dorsal bristles. Indeed, most of the Chinese specimens of *Psittacosaurus* do not preserve skin or feathers, but rich finds from a single, very strange locality have told us a great deal about their palaeobiology.

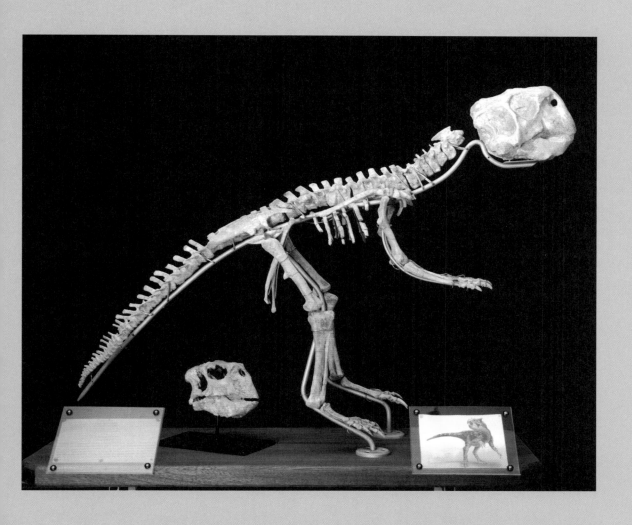

A *Psittacosaurus* skeleton mounted in a roughly life-like posture. The short arm proves that the adult *Psittacosaurus* was generally a bipedal animal, although it doubtless sank to all fours when feeding on low-growing plants.

The Chinese Pompeii

The fossil beds at Lujiatun in the western part of Liaoning Province, north-east China, are world-famous because they consist largely of volcanic rocks mixed with lake sediments. In fact, the *Psittacosaurus* specimens from Lujiatun were trapped by falling ash as volcanoes erupted around them. In 2013 we carried out an investigation at this extraordinarily rich site. Dr Chris Rogers, now a researcher at the University of Cork in Ireland and then my PhD student, had the task of logging the rock succession, identifying the environments of deposition and explaining exactly how the dinosaurs had been preserved.

It turns out that the Lujiatun beds are a special local rock unit, occurring only around the village, and part of a longer sequence of rocks laid down in ancient lakes that existed over this part of northern China for some 10 million years about halfway through the Early Cretaceous, about 123 million years ago. In most cases, the sediments are muds and silts, often containing volcanic dust, and the fossils are buried flat between these fine layers, showing wonderful detail of soft tissues in insects, fishes, lizards, small dinosaurs, birds and mammals. But at Lujiatun, all these small animals were burnt to a frazzle by the falling hot ash, and only larger animals were preserved – most commonly *Psittacosaurus*, but also the mammal *Repenomamus*, some specimens of which had partially digested baby *Psittacosaurus* inside their stomachs (this was announced in the mainstream media as 'mammal eats dinosaur', something like the apocryphal 'man bites dog' headline).

Opposite:
The fossiliferous sediments of the Lujiatun Unit (or the 'Chinese Pompeii'), the layers of volcanic ash that enclose most of the Chinese *Psittacosaurus* specimens. These are microphotographs of views under the microscope, showing different rock types. (1) Siltstone containing a partial tooth row in lower part of image and a biotite-rich volcanic rock fragment in top right of image. (2) Close-up of articulated ribs within a pink ash-like sandstone. (3) The Pink Tuffaceous Sandstone with orange coloured vitric ash. (4) Sample from the Lower Tuffaceous Siltstone, showing several pseudo-fiammé, tiny pieces of melt glass thrown out by the volcano. (5) The rock matrix of IVPP V14748, one of the museum specimens, confirming it is preserved in the same rock type as in (4). (6) Matrix of a lahar deposit, a mass of flowing ash and volcanic debris, from IVPP V14341, showing volcanic rock fragment in the lower left-hand corner of the image. These last two images (5, 6) show how geologists can match the rock matrix of museum specimens that have been purchased without provenance information to the rocks in the field.

Babies and biology

Thousands of *Psittacosaurus* specimens have been found at Lujiatun and in surrounding locations. Many of them are clutches of juvenile specimens, and every museum in China has its own clutch of *Psittacosaurus* babies. Most of these are genuine, but Chinese researchers and curators are wary of fakes – where the specimens have been glued together from different collections, or 'improved' in other ways.

My former PhD student, Dr Qi Zhao, studied microscope slides of bone from a genuine clutch, and was able to age the individuals by their growth rings, one ring being laid down each year. Of the six juveniles, five were aged about two years, and the other three years old. The older dinosaur was still hanging out with a bunch of babies – probably, like other dinosaurs, the infants kept apart from the perhaps perilously large adults – but was undergoing some major changes. *Psittacosaurus* were born as quadrupeds, and we had wondered when and how they became bipedal as adults. Qi's studies of bone histology showed that the leg bones grew faster than the arm bones, and that from a condition of four equally short limbs when they hatched, the legs elongated, and the dinosaurs became bipedal by about age three or four.

Their dietary habits, too, changed in adulthood. In a study comparing the skulls of a hatchling and an adult, visiting Italian student Damiano Landi showed that the power of their jaws increased hugely through growth, as we would expect, and that the maximum bite point shifted forwards a little in the fully grown individual, suggesting that they snipped tougher vegetation than the babies.

Opposite:
A *Psittacosaurus* nest. This specimen in the Wyoming Dinosaur Center might seem almost too good to be true. It shows about thirty babies in a clutch, with an adult skull at the right. Is this mum and her babies? More likely, this is an amalgam of many genuine fossil specimens of baby *Psittacosaurus* collected from Lujiatun, plus an isolated adult skull, combined to make a single display-type scenario.

Feathers

Most of the Lujiatun *Psittacosaurus* are just bones – all traces of their skin and feathers were burnt off by the falling hot ash. The Senckenberg specimen, first described by Gerald Mayr and colleagues in 2002, came from Sihetun, near to Lujiatun, but with fossils preserved in the more usual lake beds. Hence there are extensive traces of the skin and feathers preserved as black, carbon-rich films. The feathers along the back were described as cylindrical, tubular structures anchored deeply in the skin. There are about 100 of them, all crammed into a strip 23.5 centimetres (9¼ inches) long running down most of the length of the short back. Each bristle is about 16 centimetres (6¼ inches) long, and measures about 1 millimetre wide at the base, tapering to a point at the tip.

By 2016, Mayr was more confident that the tail bristles of *Psittacosaurus* were in fact feathers. During the intervening fourteen years, more typical feathers had been identified in other ornithischian dinosaurs, including *Tianyulong* and *Kulindadromeus* (pp. 168–77), and researchers had identified a modern analogue, the so-called 'beard' feathers of the turkey and the guineafowl. In these modern birds, the feather-like growths of the 'beard' do not emerge from follicles (pits within the skin); instead a pimple grows directly from the skin, and is covered with keratin as it lengthens. Keratin, a naturally tough, transparent protein, is the basic building material of hair, feathers and fingernails. The bristles of the turkey beard develop a longitudinal hollow as they grow, and grow in bundles, firmly fixed into the skin, just like the *Psittacosaurus* bristles.

So, they are not feathers as such, but something close to feathers. Because birds show these kinds of structures, and they occur in bunches, reflecting some developmental patterning, they are probably triggered by similar developmental programmes during growth. It's a moot point, then, whether we call them feathers, and classify them as a special type of non-branching quill that grows in certain modern birds, and likely also in *Psittacosaurus*, as a display structure.

Both male and female turkeys have the beard structures, but only males make the beard stand upright, when they are aroused and showing off to females. Perhaps *Psittacosaurus* similarly signalled to members of the opposite sex by erecting their dorsal quills, perhaps rattling them as porcupines do with their hair-derived quills, switching their tails from side to side like a windscreen wiper for maximum effect. We can imagine dozens of these dinosaurs in northern China and Mongolia 125 million years ago, flashing and shuddering their quills with a rattle of excitement during breeding season.

Opposite:
The turkey 'beard', made from a tuft of unusual feathers called mesofiloplumes that grow continuously from the skin, but without a follicle, may provide a parallel for the *Psittacosaurus* bristles. The 'beard', found in both male and female turkeys, dangles seductively from the bird's breast.

11

KULINDADROMEUS

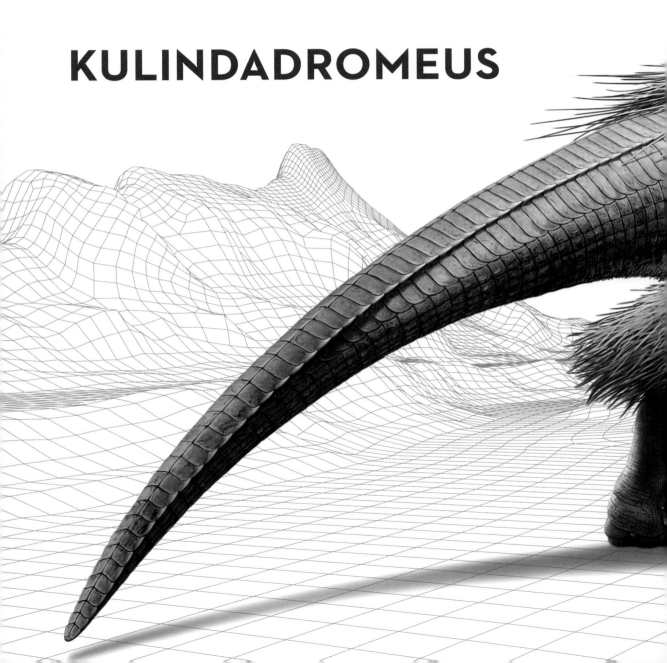

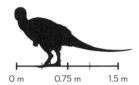

0 m   0.75 m   1.5 m

The steely stare

Siberia in the Middle Jurassic is a warmer place than we might expect; we are at a latitude of 50° North today, in line with London and Edmonton, Canada, but that translates to about 40° North in the Jurassic, closer to the Equator, and there are no ice caps. There are plenty of trees and bushes, buzzing with insect life, and in the trees you see a most peculiar-looking dinosaur; he calls to mind a wallaby or small kangaroo. His head is short, and his jaws are lined with small teeth suited to a plant-based diet.

His body is covered with short, hair-like feathers that form a soft, light-brown pelt, which changes colour as the animal moves and twists. A dark stripe runs down the middle of the back, and the light brown tints on his sides shimmer from very pale to much darker as he moves. The long tail is, unusually, scaly, lined above, below and on the sides by an angular, tube-like armoured construction. The arms and legs, too, are scaly, but with a mass of fluffy feathers on the thighs, like pantaloons. He fixes you with a steely gaze, as if daring you to make fun of his remarkable dermal appendages.

There were large predators around in the Jurassic, and the small *Kulindadromeus*, only about 1.5 metres (5 feet) long from snout to tail tip, had to be wary. We don't know these predators in detail, as only a single tooth has been found so far in the Siberian rocks. But, if it was similar to *Allosaurus*, found in rocks of the same age in North America, it would have been a hefty animal, some 9.5 metres (31 feet) in length, running on powerful hindlimbs and with massive jaws. It could easily outpace the much smaller *Kulindadromeus*, but the small feathered herbivore could perhaps have jinked from side to side, confusing the predator as its pelt flashed and rippled, alternating light and dark. *Kulindadromeus* hurtles into the shadows, and then stops still. Its variegated brown feathers and scales blend seamlessly into the background vegetation.

Are you looking at me? The discovery of *Kulindadromeus* in 2014 was an important step in understanding feather evolution. This plant-eating ornithischian dinosaur is a long way from birds in the evolutionary tree, and yet it shows a diversity of feather types. This confirmed that likely all dinosaurs had feathers of some kind.

A surprising specimen

The identification of feathers in the amazing little *Kulindadromeus* caused quite a stir. Hundreds of similar ornithopod (plant-eating) dinosaurs have been found all round the world, many of them much larger, but rarely is there any sign of skin or feathers. It seems that the Kulinda site, where the specimen was found in 2013, has yielded so much information in the exceptionally preserved fossils because of nearby volcanoes. Volcanic ash is acidic, and when it fell into the lakes and rivers in which the specimens were found, it made the waters slightly acidic, essentially pickling the flesh and skin, just as we pickle gherkins or onions in vinegar (dilute acetic acid). Acid preservation is also responsible for the exceptional preservation of the so-called 'bog bodies' of north-western Europe, human cadavers that were thrown into peat bogs, in which the humic acid released by the peat tanned their skin and preserved all the soft tissues.

The circumstances necessary to preserve these ancient feathers are a reminder of how much of these ancient beasts is lost to us. But the real importance of the discovery lies in the fact that these are ornithischian dinosaurs – members of the group that includes many key plant-eating dinosaurs such as *Iguanodon* and the hadrosaurs, forms that were, in evolutionary terms, as far from birds as possible in the dinosaur family tree. If *Kulindadromeus* truly has feathers, it could indicate that all dinosaurs had feathers, from their very origins. What was the evidence?

We were contacted by Pascal Godefroit, a professor at the Natural History Museum in Brussels, Belgium, in early 2013. At that time, Maria McNamara, now a professor at the University of Cork in Ireland, was working with me in Bristol on the preservation of dinosaur feathers and colour. Pascal had been collaborating with colleagues in Russia since 2011, and they had found the most amazing new dinosaur in Kulinda with preserved skin, scales and feathers. Could we help?

Pascal sent over some samples, and Maria was quickly able to confirm his identifications. The feathers were unusual, but that was to be expected, because this was a very distant relative of birds. The feathers Maria saw under the microscope were of three types. First were the monofilaments, single, long whiskers, from 1 to 3 centimetres (½ to 1¼ inches) long, and sprouting from the skin along the back, round the sides of the trunk and round the head. These simple whiskers are found in many dinosaurs, including commonly in *Sinosauropteryx* (pp. 28–41), but also in birds.

The second feather type was unique, a sort of plate bearing six or seven filaments. The basal plate is a scale, ranging from 2 to 4 millimetres wide, and the banner-like streamers are a fraction of a millimetre wide. These feathers are something like the down feathers of a bird, but their arrangement on a basal plate is unique. The basal plates are arranged in regular diagonal rows, covering much of the upper arm and the meaty thigh areas.

The skeleton of *Kulindadromeus*,
showing it was a typical small ornithopod
dinosaur, similar to numerous other
modest-sized plant-eaters of the Jurassic
and Cretaceous. The skull is short,
with powerful jaws for snipping plant
food. The limb sizes show it was primarily
a biped, but the arms were adapted both
for grabbing and manipulating food, as well
as for resting on the ground from time to
time. The long, stiff tail was a horizontal
balance rod when the animal ran at speed.

The third feather type is seen only on the calf of the leg of *Kulindadromeus*, consisting of bundles of six or seven ribbon-like structures up to 2 centimetres (¾ inch) long. Close inspection reveals that each of the ribbons consists of about ten very thin filaments, in a tight bunch.

As well as these three types of feathers, the *Kulindadromeus* specimen showed three types of scales. On the lower legs are small, overlapping hexagonal scales, each 3.5 millimetres (⅛ inch) long. The hands, wrists, ankles and feet are covered with even smaller, circular-shaped scales. The largest scales, the third type, occur along the length of the tail in five rows. These scales are 2 centimetres (¾ inch) wide and 1 centimetre (⅜ inch) long, and they overlap in a regular manner, the scale in front sitting partly over the tip of the scale behind, just like the slates on a roof. The directionality of overlap is important: on a roof, the slates overlap downwards, so water flows from top to bottom and does not penetrate into the house. On the tail of *Kulindadromeus*, the plate-like scales of the tail overlap from body to tail tip, so they line up with the forward movement of the animal. If they overlapped the other way, the scales might open up and catch on twigs and obstacles as the animal tried to go forward.

Cossack trousers:
Kulindadromeus is a Russian dinosaur, but conditions in Siberia in the Jurassic were warm, so he didn't need woolly trousers. However, the extraordinary fossils show a variety of small and large scales over the lower parts of the legs, and long, whiskery feathers over its thighs.

Early origin of feathers

By 2014, hints of feathers had been identified in a few ornithischian dinosaurs. First were the strange quills down the back of *Psittacosaurus*, first reported in 2002 (see pp. 154–67). Then came the curling, stiff feathers on the back of *Tianyulong*, reported in 2009. But these were both cases of unusual crests, quite different from the all-over body covering seen in birds and many small theropod dinosaurs. Up to that point, palaeontologists had been cautious about making the claim that all dinosaurs likely had feathers from their origin.

Kulindadromeus, like *Psittacosaurus* and *Tianyulong*, is an ornithischian dinosaur. Traditionally, the dinosaur group as a whole is seen as having split into two, and then three, major groups very early in their evolution, perhaps around 230 million years ago, in the Late Triassic. Theropods are all the flesh-eating dinosaurs, and they include the ancestors of birds, so it was no surprise that feathers were found in a series of theropod dinosaurs, including *Sinosauropteryx* (pp. 28–41), *Caudipteryx* (pp. 56–69) and *Microraptor* (pp. 70–83). The second group are the sauropodomorphs, including giant plant-eaters such as *Diplodocus*, *Brontosaurus*, and *Saltasaurus* (pp. 140–53) and these have never yielded feathers. Then there are the ornithischians, also herbivores, and including unarmoured forms such as the ornithopods, but also the armoured ankylosaurs and stegosaurs, and heavy-headed ceratopsians such as *Psittacosaurus* (pp. 154–67).

Our work on *Kulindadromeus* proved that ornithischians had feathers, and that many of them were just the same as theropod and bird feathers. Some, such as the banner feathers, were novel, and not known in any theropod or bird, but that's fine. Over hundreds of millions of years, it's not unexpected that feather types might diversify beyond what we still see in living birds.

Some researchers suggested that feathers might have arisen independently at least twice, once in theropods (leading to birds) and once in ornithischians, but the simpler view is that they originated once only, at the point of origin of dinosaurs, or even lower in the family tree (see p. 214). Our study of *Kulindadromeus* showed that there are several shared feather types between the groups, and their absence in most dinosaurs might well simply be because of incomplete preservation. Therefore, perhaps even the first, Triassic, dinosaurs had feathers. At that time, feathers would have had nothing to do with flight, and perhaps more to do with the fact that most, or all, dinosaurs seem to have been warm-blooded, and feathers, in small animals like *Kulindadromeus* at least, provide insulation.

Those scales

The scaly arms, legs and tail of *Kulindadromeus* are intriguing. Are these simply the primitive scales of the reptilian ancestor of dinosaurs, or are they something else? Among modern animals, rats have scaly tails and chickens have scaly legs. Genetic evidence tells us that these are not primitive holdovers, but that these particular tail and leg scales evolved over the millennia from feathers.

It seems most likely that ancestors of *Kulindadromeus* had feathery arms, legs and tails, but natural selection caused these feathers to modify themselves to specialist scales. So, the first dinosaurs might have been covered all over with feathers, and this condition continues in many descendants. But in some, such as *Kulindadromeus*, feathers were modified to scales, perhaps to enable faster running (feathery legs might interfere with each other, just as sprinters prefer not to wear baggy trousers when they are running) or to enable the animal to lose body heat and avoid becoming overheated; perhaps to act as a signal, or for some other reason entirely.

The discovery of the amazing little dinosaur *Kulindadromeus* in 2014 has opened up a whole new world in the study of dinosaur feathers. No longer can we see feathers as unique to birds, or even to the direct ancestors of birds, but perhaps to all dinosaurs or, even more broadly, to dinosaurs and their cousins, the pterosaurs (see pp. 212–14). When we began studying dinosaur feathers in about 2005, we had no idea where the work would lead!

Four close-up views of the *Kulindadromeus* specimens, showing the different feathers and scales in different parts of the body. (1) Partial arm, shown by orange-stained bones and cavities, with small, round and lizard-like scales, each about half a millimetre across. (2) Some specialized feathers, called 'banner feathers', composed of long, strand-like filaments extending from scale-like bases that are arranged in regular rows, from the thigh region of the leg. (3) Impressions of two scale types, larger and angular scales to the right and smaller circular scales to the left, around the shin region of the leg, showing also the tibia bone. (4) Fragments of skin showing numerous small, circular scales between the ribs (shown as orange-stained structures).

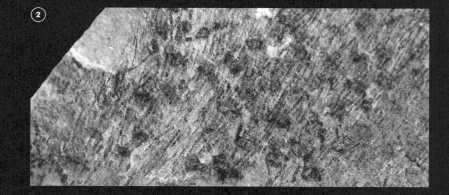

③

④

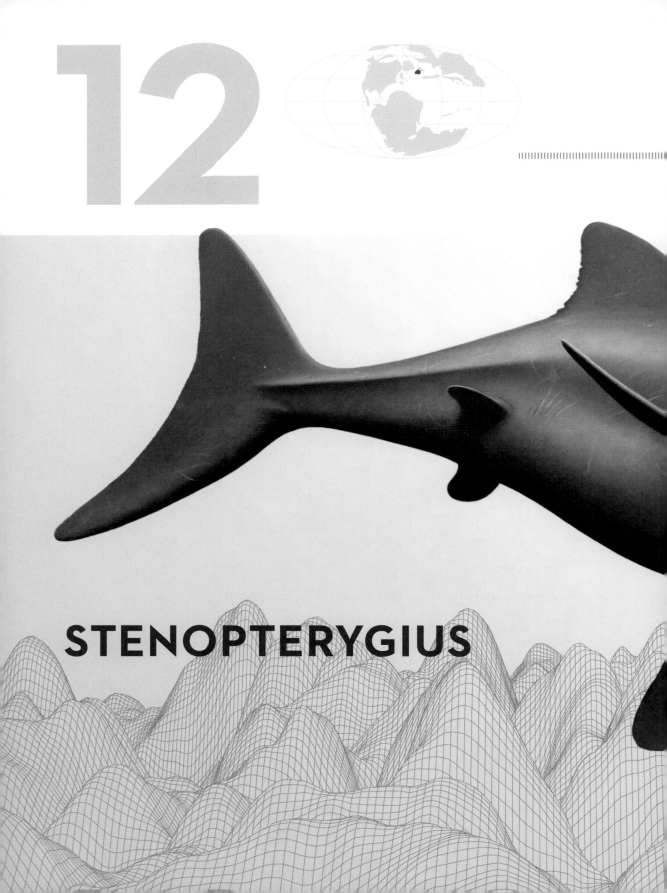

12

STENOPTERYGIUS

Foiled again

This is a very different scene: we are underwater, swimming with the ichthyosaur *Stenopterygius*. This is not a dinosaur, but a marine reptile that lived in the Mesozoic seas, alongside a whole array of sea monsters that evolved in the Triassic and variously dominated food chains around the world, feeding on fishes or cephalopods. There were even some giant marine reptiles that might have preyed on our *Stenopterygius*.

As we watch, *Stenopterygius* swoops after a belemnite, an extinct relative of modern squid and octopus. This belemnite has a fleshy body and fins and it swims backwards, just as modern cephalopods do. We know, too, that it has an ink sac, and so, like its modern relatives, likely squirts ink when alarmed, and zips off by blasting jets of water through its siphons. By the time the predator has recovered and snapped a few times at the ink cloud, the belemnite has long disappeared to safety.

The *Stenopterygius* is not too fazed, as this is not an infrequent occurrence, and he lines up to chase another group of belemnites. *Stenopterygius* is typical of many ichthyosaurs from the Early Jurassic, measuring 2–3 metres (6½–10 feet) in length and built for maximum hydrodynamic efficiency. In fact, the ichthyosaur is shaped like a torpedo: nearly circular in cross section, with a long, slender snout, and fins and paddles that curve back from the body. This classic hydrodynamic shape evolved independently also in other fast swimmers, such as sharks, tuna and dolphins.

Stenopterygius is all black, except perhaps for a dark grey belly. This dark belly is unexpected. Most fast-swimming fishes and marine tetrapods have a white belly, an example of countershading, which allows the animal to blend in with the colour of the sea water whether it is viewed from above or below. From above, a predator would be staring into the inky depths of the ocean, and a black-backed animal, even a large one, would perhaps seem to be invisible. From below, a white-bellied form would blend with the bright light of the sky.

A black belly is quite a claim. Can we be sure? And if *Stenopterygius* did have a black belly, what does it mean?

Opposite:
Three-dimensionally preserved skeleton of the Early Jurassic ichthyosaur *Stenopterygius* with the skull, vertebrae, shoulder blades, sternum and front swimming paddles. The skull clearly shows the large eyes, capable of picking up movements in deep water where light is limited, as well as the long jaws lined with sharp teeth, perfect for snatching and trapping fishes and cephalopods.

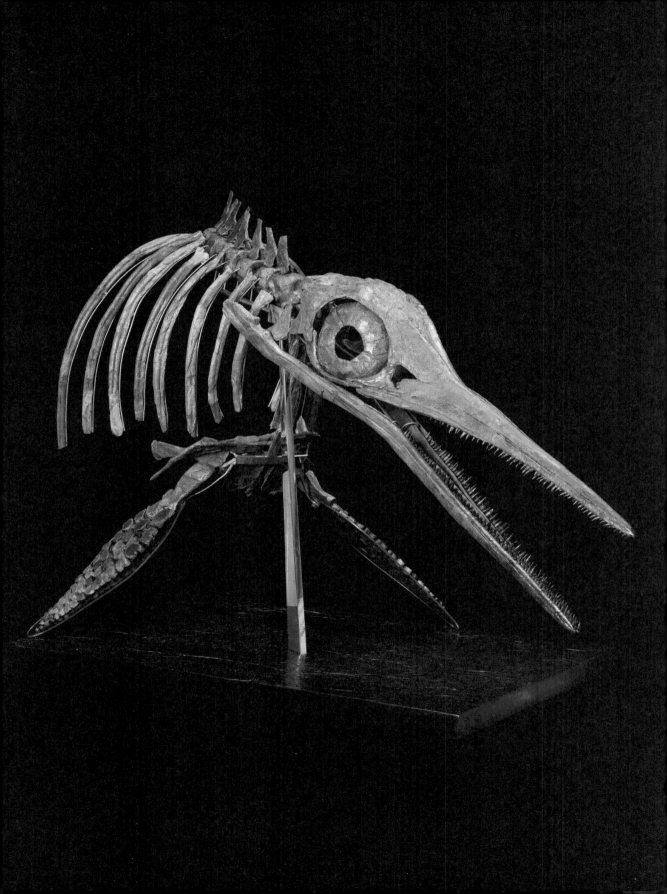

Ichthyosaur palaeobiology

Ichthyosaurs were first discovered about 1800, and indeed specimens might have been excavated before then. In certain rocks of the Triassic and Jurassic in Europe, fossils are plentiful, and they are often complete. Ichthyosaurs were entirely marine reptiles, and we now know they arose in the Early Triassic from land-dwelling ancestors. Just as the terrestrial ancestors of whales adapted to life in the sea, a number of different lineages spawned marine predatory forms in the Early Triassic; this was a time of great environmental turmoil following the end-Permian mass extinction, the greatest crisis in the history of the Earth and of life.

These early marine reptiles preyed variously on fishes, lobsters and molluscs. Some became adapted to feeding on hard-shelled prey by evolving pavements of hard, flat-topped teeth with which they ground up the shells of oysters and other shelled creatures that lived around the coasts. The fish-eaters, like the ichthyosaurs, mainly evolved long, narrow snouts lined with numerous sharp teeth. Such a snout is also seen in some crocodiles today; as they snap their jaws shut, their long teeth make a kind of cage that catches slippery fish or squid. As the jaws close completely, water is squirted out sideways and the wriggling fish, sometimes impaled on some teeth, is firmly gulped down. There is no chance of escape.

We have reconstructed the diets of many ichthyosaurs from their gut contents and coprolites (fossil excrement), as well as finds of unfortunate prey animals. Gut contents in particular cannot be denied. Several Early Jurassic ichthyosaur specimens, including *Stenopterygius*, have been found with masses of belemnite hooklets inside their rib cages. These are tiny, hook-shaped fossils, each about 1 millimetre long, and in life they lined the tentacles of the belemnite and allowed it to grip and pass its prey towards its mouth, which was located in the middle of the crown of tentacles. Such hooklets seem to have accumulated inside the guts of ichthyosaurs; either they ate simply huge quantities of belemnites each day, or perhaps the hooklets built up over time and for some reason were not excreted. The hooklets are made of chitin, a polysaccharide similar chemically to cellulose, which evidently did not break down in the stomach acids of the ichthyosaur. Most marine predators, such as crocodiles and killer whales, have stomach acids that can dissolve the bones and shells of their prey, so very little remains after they have digested their food, and this was likely the same for ichthyosaurs.

Coprolites may show a faintly spiral pattern, perhaps matching the final portion of the gut, but it is often hard to determine which coprolite belongs to which maker. Even more unusual is that several examples of ichthyosaur vomit have been identified, often consisting of hundreds of belemnite guards. The guard is the internal shell of the belemnite, much larger than the hooklets. Evidently ichthyosaurs could not crunch up the guards and spit them out, and instead swallowed the animals whole, digested the fleshy body and tentacles, and then honked up the guards, sometimes a hundred or more at a time. This suggests that some larger ichthyosaurs at least were feeding on shoals of swimming belemnites, perhaps, as some larger whales do today, forcing the cephalopods to form a tight shoal by swimming around them and then engulfing a whole mass of them, ignoring their ink clouds. *Stenopterygius*, however, might have been too small to engage in this kind of mass slaughter, unless groups of three or four cooperated in herding the prey animals.

Leaving land behind

The first ichthyosaurs that crept into the ocean could probably still walk on land, but their skeletons are yet to be found. We have to look to the better-documented evolution of whales to understand how this transition might have worked: early whales lived their lives in two phases, diving into the water to catch their food, as seals and otters do, and crawling out on land to have their babies. However, all the ichthyosaurs that are known from fossils are fully adapted for swimming, with paddles and their hydrodynamic, bullet-like body shape. Traces of the ancestral fingers are there – the earliest ichthyosaurs still have the standard five fingers of land vertebrates – but enclosed in thick skin like a great mitten.

Ichthyosaurs powered through the water by beating their bodies from side to side, using their front paddles mainly for steering. If the ichthyosaur was moving forward at some speed and saw an obstacle, it would stick out its left fin, forcing its head and body to turn right. The earliest ichthyosaurs were more serpentine, whereas *Stenopterygius* and all later forms were relatively shorter, more tuna-like in shape, with a fat body and broader tail fin that generated their power. In careful experimental studies of swimming, my PhD student Susana Gutarra has shown that ichthyosaurs swam as efficiently as any modern shark or dolphin, and they had this efficiency from the start.

Opposite:
Slab of limestone with fossil belemnites and four ichthyosaur vertebrae across the top. Probably these fossils are preserved together by chance, but they remind us that Early Jurassic ichthyosaurs fed on belemnites as a major part of their diet, and then had to spit out the calcite guards inside the belemnites, forming so-called 'vomitite' accumulations.

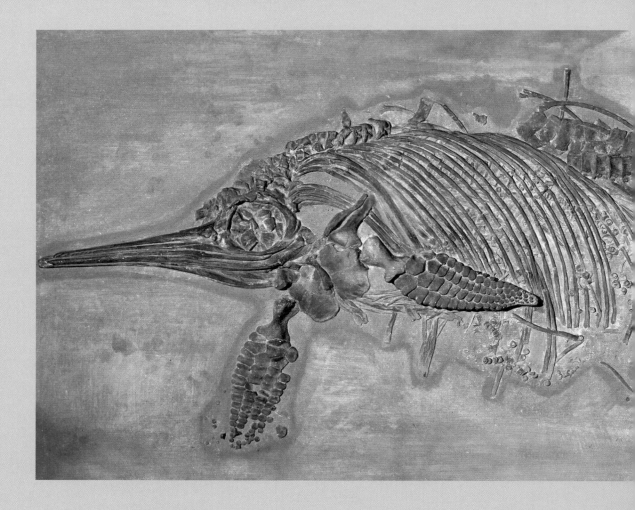

Indeed, for all the changes in their body shapes, swimming efficiency remained rather constant, although the tuna-like body shape enabled ichthyosaurs to become larger and larger (*Stenopterygius* was between 3 and 4 metres, or 10 and 13 feet, in length), and to improve their efficiency by reducing the relative amount of drag.

The earliest forms may have struggled on to land to produce their babies, but *Stenopterygius* definitely did not. We know that they gave birth to live young at sea, just as modern sharks and dolphins do, because numerous specimens have been found bearing foetuses. Many of these come from a locality in southern Germany called Holzmaden. These have been studied closely, and the small ichthyosaurs had not been eaten by cannibal adults, but are aligned in groups matching the likely locations of the paired ovaries of the mother. Female adults might carry as many as six or eight babies. There are even rare examples of a pregnant adult with one or more babies outside the body. Did the mother and her babies die during birth? Probably

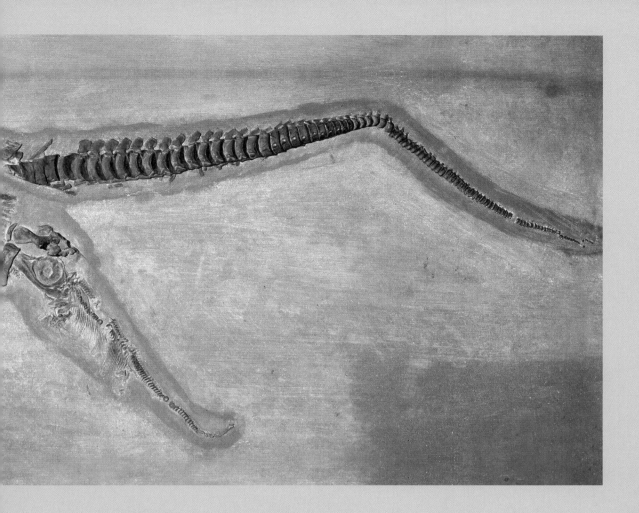

A pregnant ichthyosaur (*Stenopterygius quadriscissus*) from the famous ichthyosaur deposits of the Early Jurassic at Holzmaden, southern Germany. This specimen shows several developing babies inside the rib cage, as well as one example apparently in the process of being born.

not, because they would probably not then all be found so close together. It's assumed the unfortunate female died at an advanced stage of pregnancy, perhaps because she was carrying too many embryos, and the babies found outside her body exploded out through generation of decay gases in her body. Not such a touching scenario!

The babies are consistently found lined up to be born tail-first, not head-first, as in nearly all cases of live birth from humans to lizards – but not whales. By being born tail-first, the baby whale gains crucial time to rush to the surface to take its first gulp of air. If it was born head-first, it might very well drown before it had entirely wriggled free from its mother.

An amazing historical document.
This drawing and inscription were made
by Elizabeth Philpot in 1833: she shows
an example of an ichthyosaur skull, drawn
using the ink (sepiomelanin) from a fossil
squid of the same age as the ichthyosaur.

Black skin and black ink

Belemnite ink, like the ink of modern cephalopods, was made of a special kind of melanin called sepiomelanin, and is as black today as the day it was produced. Elizabeth Philpot (1780–1857) made a drawing of one of the remarkable ichthyosaur specimens collected by Mary Anning (1799–1847), the famous fossil collector of Lyme Regis; she scraped some fossil melanin from a fossil belemnite, worked it into a paste and used the Jurassic ink to make a drawing of an ichthyosaur. It was his studies of fossil sepiomelanin that led Jakob Vinther to realize that fossil melanin can be identified by the sausage-shaped microbodies that others had mistakenly thought to be bacteria (see pp. 50–55).

A rather forgotten paper published in 1956 produced the first evidence that ichthyosaurs were black. Mary Whitear (1924–2018), a marine biologist specializing in the structure and function of fish skin, described a piece of exceptionally preserved ichthyosaur skin from the Early Jurassic of the Dorset coast in southern England. She noted its overall black colour, but also, crucially, she observed melanocytes, the cells in the skin that produce melanin. She reported that the melanocytes were dendritic in shape, that is like the branches of a tree pointing out around the centre, and that the cells contained 'granules of a reddish brown pigment' that she interpreted as dried-up melanin.

More detailed investigation had to wait. Nearly sixty years after this amazingly prescient paper was published, Johan Lindgren and colleagues from Lund University in Sweden looked for ichthyosaur melanosomes in a 2014 study. They took samples from a *Stenopterygius* specimen from Yorkshire and found abundant eumelanosomes all over the skin. By comparison with *Stenopterygius* specimens from Holzmaden, they inferred that the black colour extended all over the body. This is because the Holzmaden specimens often show a black-coloured body outline all round, even confirming the all-encompassing skin mittens over the paddles and the dorsal fin on the back,

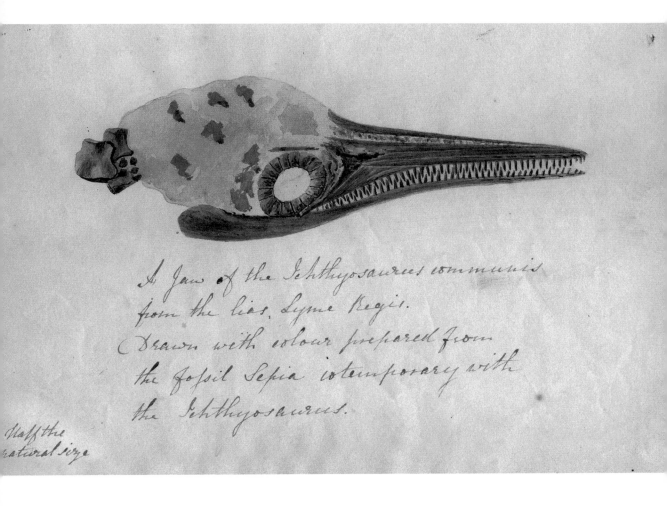

A Jaw of the Ichthyosaurus communis
from the lias, Lyme Regis.
Drawn with colour prepared from
the fossil Sepia cotemporary with
the Ichthyosaurus.

Half the
natural size

which contains no bony skeleton and, as with the dorsal fins of dolphins, was entirely built from skin.

Chemical analyses of ichthyosaur skin samples showed they were rich in carbon, and Time of Flight Secondary Ion Mass Spectrometry analyses (see pp. 102–5) confirmed that a key chemical constituent was eumelanin (black or brown pigment), while Lindgren found no evidence for phaeomelanin (red or yellow). Therefore, the researchers assumed a black colour, and black all over, because they could not identify areas around the body that apparently lacked melanosomes.

A note of caution was sounded by Maria McNamara from the University of Cork in 2018, who reminded researchers that melanosomes in fossils do not all derive from skin or feathers. In fact, many internal organs, such as the liver and the spleen, contain abundant eumelanosomes, and during fossilization these can migrate to sit around the body, for example in the black integumentary rim that outlines the Holzmaden ichthyosaur skeletons. The non-integumentary melanosomes differ in shape and packing structure from integumentary melanosomes, and further study of the ichthyosaurs is required to determine whether they do indeed represent cells from the skin. However, the uniform appearance of the black organic rim all round the body does seem to suggest that the black colour was uniformly distributed around the whole body.

If some ichthyosaurs at least, such as *Stenopterygius*, did not have a pale counter-shaded belly, what does that mean? One suggestion is that they lived in very deep water, far below the levels penetrated by sunlight,

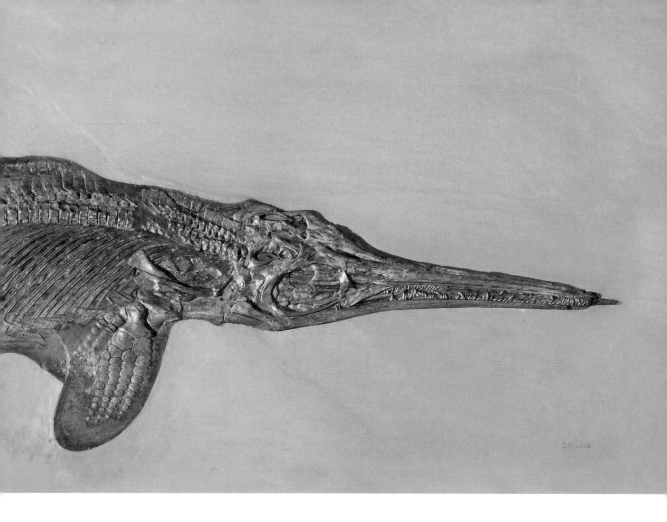

A further specimen of the Early Jurassic
ichthyosaur *Stenopterygius* from Holzmaden
in southern Germany. This example is preserved in
a flattened condition, but there are clear traces of
the skin, full of melanin, indicating a generally all-
over black colour in life. The exceptional preservation
of soft tissues confirms the shape of the upper part
of the tail and the dorsal fin, neither of which could
be seen in the skeleton.

and so countershading would have served no purpose. At these depths,
all creatures are creeping around in the gloom, and organisms may evolve
structures that flash phosphorescent colours to warn off predators or lure in
prey, or may simply have huge eyes, as the ichthyosaurs do, to pick up what
faint light penetrates to the depths.

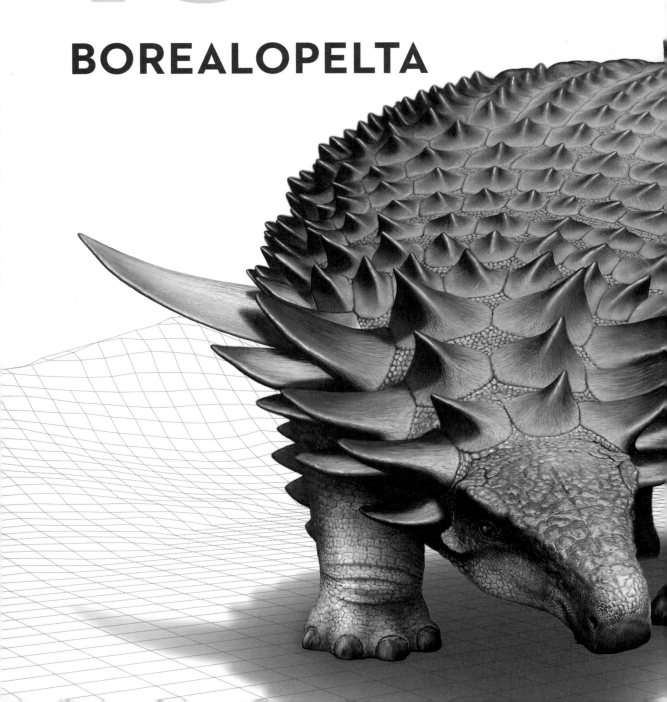

13

BOREALOPELTA

0 m   5 m

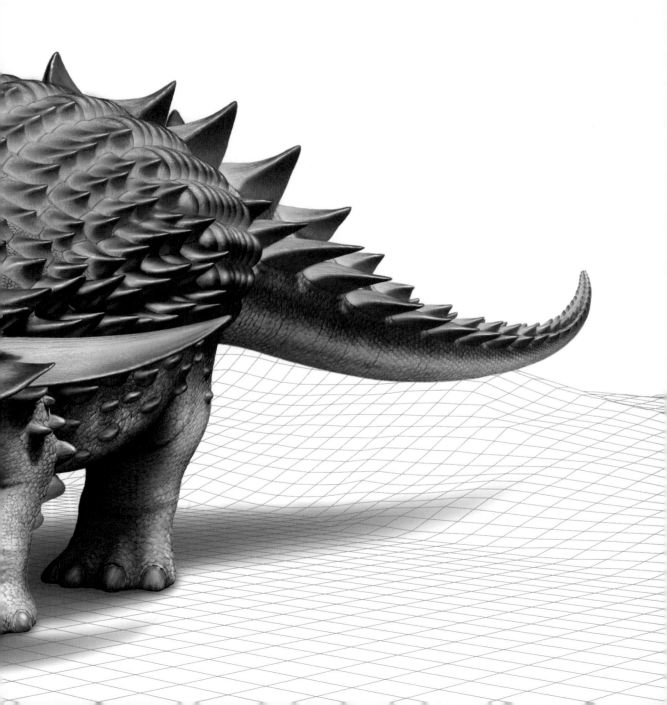

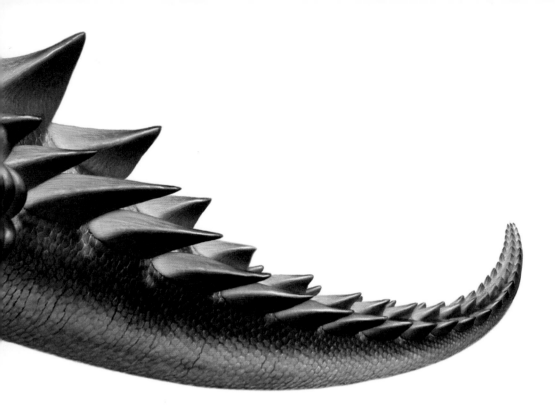

Borealopelta had a short but powerful tail that it could switch from side to side. Any predator would have to step carefully.

Spines and more spines

We are in the Early Cretaceous in Alberta, Canada, some 120 million years ago. In the distance we see a reddish-brown animal that looks like a spiky turtle. It ambles towards us, becoming larger and larger, and we realize perspective was playing tricks on us. This is no turtle. As he comes closer, we see this is a monster, an animal more than 5 metres (16 feet) long, and easily weighing over a tonne. There is no living animal this size and shape; it's more like a tank.

The head is broad and the snout short, with a blunt end. The eyes are small, located on the side of the head and entirely surrounded by bony, armoured plates. Round the side of the head and down the short neck are short bony spines, each ending in a sharp point. These continue in rows down the beast's back and sides. The body is broad and low, and there are as many as twenty rows of spines running the length of the torso. Larger spines rise from the neck and the back of the skull; two are especially threatening, each half a metre (1½ feet) long, emerging from the shoulders. These shoulder spines are shaped like powerful medieval broadswords, although probably not used for killing. The tail is also encased in multiple spine rows, running right to the diminutive end.

The skin underneath the neck and belly does not carry spines, but scales. And whereas the back is a reddish-brown colour, the belly is pale.

This is *Borealopelta*. As he walks forward, his tail swings from side to side: a formidable weapon. His eyes, too, range from side to side, and suddenly he stops; he has spotted a group of five *Deinonychus* swaggering and chittering among the trees. The *Borealopelta* hunkers down, protecting the soft underbelly beneath his menacing array of spines. The *Deinonychus* rush up and swipe at him with the sharp sickle claws on their toes. If they just could reach his belly skin, they could rip through, and expose the internal organs through a metre-long gash. But they know they are wasting their time. This herbivore is so huge he would feed ten *Deinonychus* for a week, but there is no weakness they can exploit. They play around his sides, one more daring *Deinonychus* leaping on his back and dancing through the forest of spikes. The *Borealopelta* shrugs him off, with an irritated grunt. The young *Deinonychus* falls heavily, tearing his leg on one of the *Borealopelta* side spines. He limps off, chastened.

An astonishing discovery

Borealopelta was announced in 2017, and the new specimen was immediately hailed as the best-preserved large dinosaur ever discovered. As we have seen, there are many remarkably well-preserved smaller dinosaurs, from China for example, but the condition of this 1-tonne giant was without precedent. The skeleton was found in the Clearwater Formation of the Suncor Millennium Mine at Fort McMurray, Alberta. This rock formation is largely marine in origin, and had previously yielded skeletons of marine reptiles such as ichthyosaurs and plesiosaurs. So to find a dinosaur, and a giant one at that, was quite a surprise.

The Suncor Millennium Mine is one of many in northern Alberta that exploits the Athabasca oil sands (or tar sands), a huge geological deposit covering over 100,000 square kilometres (38,600 square miles) of the province and containing an estimated 1.7 trillion barrels of bitumen that are extracted for conversion to oil, yielding a huge income for the Province (some of which gets recycled back into dinosaur study and dinosaur museums).

The *Borealopelta* skeleton was uncovered in 2011, in the course of this mining activity. Alberta law states that all dinosaur fossils belong to the Province and must be reported to the Royal Tyrrell Museum of Palaeontology, and so Suncor alerted the museum. When Royal Tyrrell palaeontologists Don Henderson (a former PhD student of mine in Bristol) and Darren Tanke visited the site, they confirmed it was a skeleton and worth excavating,

Following pages:
The Suncor Mine, northern Alberta, a huge opencast mine from which rich oil reserves are extracted from the Athabasca tar sands. Dinosaurs are only rarely recovered from this site.

but at first assumed it would be another plesiosaur or ichthyosaur. They were astonished when, on closer study, they saw it was a dinosaur. How could such a large land animal have ended up hundreds of miles offshore, to be fossilized in a marine rock deposit?

The fossil was marooned 8 metres (26 feet) up a high cliff, and it took fourteen days to extract it from the rock. There were hair-raising moments, such as when the huge block containing the skeleton was being lifted free by a crane and it broke in two. The broken block was lowered to the ground and the museum staff wrapped all of the excavated blocks in thick sackcloth dipped in plaster, bandaging up every angle for protection, just as nurses bandage and plaster broken arms and legs in hospital. Then museum technician Mark Mitchell spent six years removing all the rock from the fossil; the dinosaur was named *Borealopelta markmitchelli* in his honour.

The skeleton alone was a spectacular addition to the collections of the Royal Tyrrell Museum. After all, it is the only dinosaur known from those marine sediments and, perhaps more importantly, it is Early Cretaceous in age, whereas nearly all the other Alberta dinosaurs are Late Cretaceous. So, *Borealopelta* fills a long gap in dinosaurian evolution in North America.

But it was the soft tissue preservation that amazed the scientists, and this contributed to the fact it took six years to extract the beast from the rock. At all stages, Mitchell carefully checked the sediment around the bones for traces of soft tissues such as skin and internal organs. Remarkably, the whole carcass is also undistorted, retaining more or less its original three-dimensional shape.

Opposite:
The type specimen of *Borealopelta*, seen from the side (1) and from above (2). In the upper image, the head is looking at you, and the flattened armour-bearing carapace extends behind. The lower image shows the different shades of colour along the lines of armoured bony plates.

Skin, spikes and colour

In their announcement of the specimen in 2017, a team of seven authors, led by Caleb Brown of the Royal Tyrrell Museum, and including Don Henderson, presented a detailed account of every bone, osteoderm (bony scute or plate set into the skin), and soft tissue fragment, together with chemical analyses and other studies of these tissues. They came to some startling conclusions.

Exceptional preservation of the back of the dinosaur meant that not only were the bony scutes or osteoderms preserved, but also traces of the keratin sheaths that covered them and the skin that held them all together. These skin samples were tested by Time of Flight Secondary Ion Mass Spectrometry (see pp. 102–5), which confirmed the presence of abundant melanin in the skin, and in fact it was melanin with added sulphur in the form of the organic chemical benzothiazole, a clear indicator that all the melanin was phaeomelanin. As we saw earlier, phaeomelanin gives ginger colours in bird feathers and human hair (see p. 52), and so the phaeomelanin in the skin of *Borealopelta* likely gave the skin around and between its dorsal spines a reddish tinge.

Each of the spines was covered by a keratin sheath, like the horns of cattle and deer. The protein keratin is the base material of our fingernails, as well as of hairs, feathers and reptile scales. The keratin horn sheaths in *Borealopelta* were finely grooved and heavily pigmented with the reddish phaeomelanin also detected in the skin. So, this was a sandy-red dinosaur all over.

Not only that; perhaps the long shoulder spines were a different colour from the others, and even phosphorescent. Photographs taken under ultraviolet light showed that the tip of the spine fluoresced; whether this is some accident of preservation or chemical alteration, or a trace of something original is hard to say. But could this dinosaur have signalled others at night with flashing spine tips? Similar fluorescent effects have also been reported on the crest of the pterosaur *Tupandactylus* (pp. 220–33).

Borealopelta also appears to have been countershaded, just as *Psittacosaurus* (pp. 154–67) and *Sinosauropteryx* (pp. 28–41) were. *Borealopelta*'s belly was less well preserved than its back, because it fell on its back into the sediment, but enough is preserved to show that there is no trace of melanin in the belly area. So Jakob Vinther, who had also studied countershading in *Psittacosaurus*, was able to suggest that the countershading line, namely the division between ginger-red above and pale below, occurred just below the lowest occurring spines on its sides.

The discovery that this 1.3-tonne giant was countershaded was a surprise. Among modern mammals, it's only the smaller ones – those that are threatened by predators – that show countershading. This is because they need camouflage to help them hide in plain view, becoming two-dimensional in appearance to the approaching predator. Any modern mammal that is approaching half a tonne, including moose, rhinoceroses and elephants, shows plain colour all over its body, with no hint of a pale belly. This is because they are large enough to escape predation pressure. The countershading in *Borealopelta*, then, indicates that it was harassed by predators – not the arrogant little *Deinonychus*, but great theropod dinosaurs like *Allosaurus*. Countershading, then, might have helped *Borealopelta* to blend into a background of red-brown soil and rock.

Preservation

The greatest mystery of the *Borealopelta* specimen is how it came to be so well preserved. It was lying on its back in the rock, legs pointing straight up in the air, and this was how it must have landed on the seabed. It's likely the animal died on land, perhaps on the coastline. During the Cretaceous, sea levels were higher than today, and North America was bisected by the Western Interior Seaway, running north and south through Alberta and down to Texas. Perhaps there was a storm, or an unusually high tide, and the dinosaur carcass was lifted from the shore and floated out to sea. Likely for a few days it was bloated with decomposition gases, and so it would have floated belly-up, as is often observed when dead horses or cows are washed away by floods. After some days, the belly perhaps became so distended that it burst, releasing huge volumes of foul gas, and suddenly causing the heavy skeleton and skin to sink to the seabed, where it landed on its back and became securely lodged in place thanks to its many spines.

It hit the seabed with some force; the underlying layers of sediment were distorted by the impact and weight of the carcass. Normally such a large animal's appearance on the seabed would attract all kinds of scavengers, who would feast for days, stripping the carcass to the bone, but apparently our *Borealopelta* escaped this fate. Over a short time, the whole carcass became locked solid in a siderite concretion. Siderite is iron carbonate that forms today at the bottom of lakes and seas where oxygen is absent at depth.

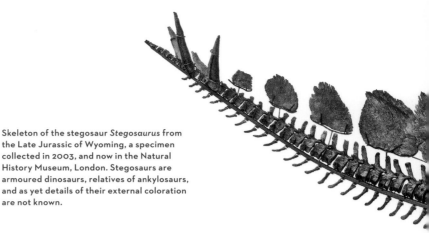

Skeleton of the stegosaur *Stegosaurus* from the Late Jurassic of Wyoming, a specimen collected in 2003, and now in the Natural History Museum, London. Stegosaurs are armoured dinosaurs, relatives of ankylosaurs, and as yet details of their external coloration are not known.

Whether the seabed at the site of Suncor Mine was anoxic overall, or somehow the chemistry of the decaying carcass made it anoxic, is not clear; either way, the skeleton became encased in a tough and impermeable iron-rich armour. This shell shows traces of cracks, which perhaps formed as the carcass collapsed and the remaining bodily fluids were expelled as sediment accumulated above, compressing the belly. It's this impact on the seabed and instant armouring in a tomb of iron-rich siderite that seems to have preserved the spines and skin intact.

Armoured dinosaurs

While palaeontologists had excavated armoured dinosaurs since the 1830s, when the first ankylosaur, *Hylaeosaurus*, was recognized in the Early Cretaceous of southern England, nothing like this specimen had ever been found before. There are two groups of armoured dinosaurs, both closely related to each other: the stegosaurs and ankylosaurs.

Stegosaurs, most famously *Stegosaurus* from the Late Jurassic of North America, have bony plates and spines arranged down the middle of their back, and sometimes an odd spine over their shoulder region. But no such remarkable specimen of a stegosaur has shown us information about their colour patterning. Among ankylosaurs, also known from the Jurassic and Cretaceous, *Borealopelta* is reasonably typical in terms of size and armament. There are two ankylosaur groups: the nodosaurids, such as *Borealopelta* and *Hylaeosaurus*, and the ankylosaurids, which famously had great bone clubs at the ends of their tails. In the classic scenario, the Late Cretaceous North American ankylosaur *Euoplocephalus* squares up to a menacing *Tyrannosaurus* rex and gives it a great disabling thump in its belly using the club at the end of its tail. Whether it could break a *T. rex* leg is yet to be determined, but all the ankylosaurs, including *Borealopelta*, were formidable opponents even for the top predators of the age.

In 2020, something more was discovered about *Borealopelta* – his last meal. Inside the stomach area, palaeobotanists identified spores and fragments of ferns, cycads and conifers. Ankylosaurs had small teeth, and could probably not have reached very high, so we envisage them moving slowly across the landscape, crunching up all the plant matter they can find – leaves, fruits and wood. Our *Borealopelta* was quite choosy, and most of the food belonged to a particular kind of fern. There was also charcoal, so perhaps the landscape had been burned. The fern sporangia, or fruiting bodies, in his stomach, would be open in high summer, so likely this *Borealopelta* died in July.

Following pages:
Gut contents of *Borealopelta*, showing numerous stones the animal had swallowed to help it grind up its food. The black material consists of organic remains of plants it had eaten.

14

ANUROGNATHUS

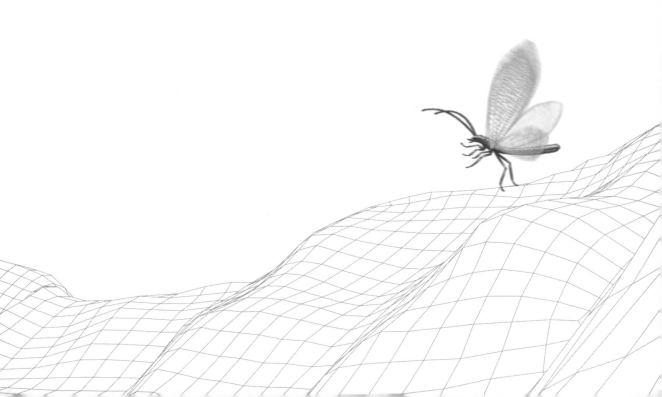

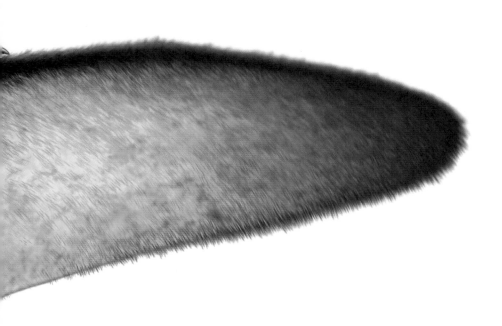

Insect killer

We now travel back to the Middle to Late Jurassic of northern China, some 165 million years ago, when the multi-coloured, feathered *Anchiornis* (pp. 42–55) was strutting and swooping in the lush forests. Also flitting through the treetops is a much less showy animal – but a real flyer. This is *Anurognathus*, a pterosaur (flying reptile). With its superior vision, it spots a lacewing flapping lazily over a pond, and dives. The insect feels a rush of air as the anurognathid approaches, but is trapped between the beating wings, which funnel streams of air inwards; it becomes entangled in long whiskers, and is engulfed and swallowed in a second. End of fly.

The predator is a furry, brown animal, with a wingspan of only 35 centimetres (13¾ inches) – about half the size of a pigeon – but it flies with skill, skipping and cruising between the tree branches. It can operate each wing independently as it approaches an obstacle or takes a turn, tucking them into its body to ensure they are not damaged in this densely packed environment. Its short, deep wings are similar to those of modern forest-dwelling birds, whereas birds that live in predominantly open terrain, like gulls, have long wings, and so must avoid complex landscapes where their wings might be damaged.

The wings of the anurognathids were fluffy and brownish in colour. The hand was used to grasp prey.

Opposite:
The anurognathid *Anurognathus ammoni* from the Lithographic Limestone of Solnhofen, Bavaria, southern Germany, the site of discovery of *Archaeopteryx* (see pp. 86–87). This beautifully complete fossil shows the long legs, short tail, long, powerful wings and broad-jawed skull.

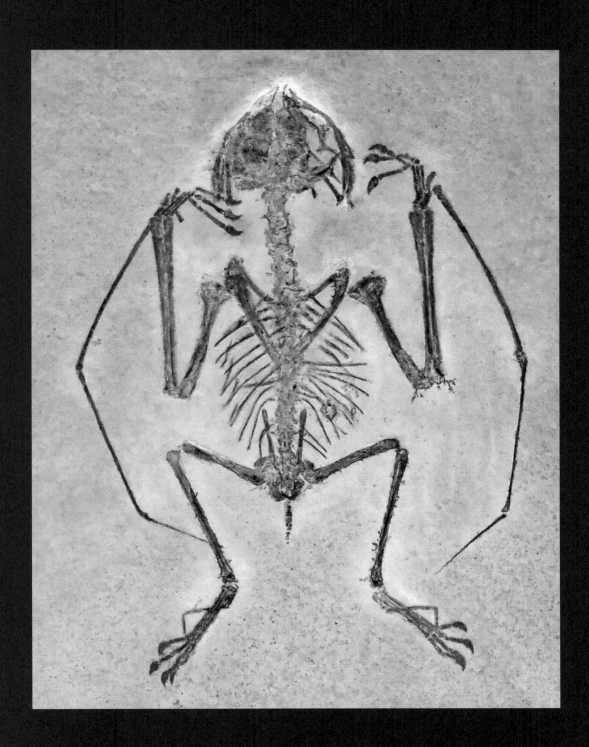

The Vanuatu flying fox, a large bat that feeds on fruit and other plant food. Unlike its more famous cousins, the nocturnal microbats, the flying foxes forage by day.

Our anurognathid was discovered in 2017 as part of the Yanliao fauna from Chinese Inner Mongolia, but has not yet been named. The anurognathids are a small group of Jurassic and Early Cretaceous pterosaurs that differ from all other pterosaurs in their short faces. The first anurognathid, *Anurognathus*, was reported from the Late Jurassic of Germany in 1923, and since then several species have been identified from superb specimens from China, Kazakhstan, Mongolia and even the United States.

The new Chinese anurognathid has a short face and broad mouth. The eyes are large, and they face forward, which means that this animal had binocular, or three-dimensional, vision. Many animals, such as dogs, cattle and horses, have separate fields of vision for each eye. But flying animals today, such as birds and bats, have swivelled their eyes to the front of their face – indeed, as did our monkey ancestors. This means that the fields of vision of the left and right eyes overlap, and it's this overlap that enables us to see in three dimensions and to judge distances. It's important these animals are able to see in three dimensions, so that they can leap from branch to branch or come into land securely. If a monkey had eyes on the sides of its head, like a dog does, it would sail past its target and land in a heap. The large eyes of our anurognathid also suggest that it had excellent vision, and was perhaps able to pick up prey movements in dim forest light, and at some distance.

Its body is entirely covered with short, light brown fur, with some variation in colour over the back. This fur was made up of tiny whiskers, some of them branching, likely simple feathers of some kind. The whiskers are dense over the head and body, providing insulation for a high blood temperature and high-activity lifestyle. The whiskers are more thinly distributed over the wings, and the hands and feet are more or less bare.

Pterosaurs

The pterosaurs are a fascinating group of flying reptiles, whose span of time on the Earth almost exactly matches that of the dinosaurs. The oldest specimens come from the Late Triassic, especially from northern Italy and Switzerland, and they occur in many locations worldwide throughout the Jurassic and Cretaceous. Most of the earlier forms were small, perhaps little more than half a metre (1.5 feet) in wingspan, and they mostly (except for the anurognathids) had a long tail. Later forms, mainly from the Cretaceous onwards, became much larger, reaching wingspans of 3–5 metres (10–16 feet) in the Early Cretaceous, and 5–12 metres (16–40 feet) in the Late Cretaceous. The last pterosaurs died out at the same time as the last dinosaurs, when the great asteroid hit the Earth at the end of the Cretaceous, 66 million years ago.

When pterosaurs were first discovered over 200 years ago, they were interpreted correctly as flyers – their long wings and lightweight body were clear. Some, such as the anurognathids, had short, deep wings, suited for dodging obstacles in dense forests. Others had long, slender wings for soaring; others still were adapted for speed. Palaeontologists have been helped in their studies by aeronautical engineers, and there is now quite a good understanding of how the larger pterosaurs operated (see pp. 226–28).

Some of the early finds from Germany showed hints of a kind of pelt, and so palaeontologists began to reconstruct the pterosaurs as hairy, bat-like creatures in the 1840s. Even in those days biologists understood that actively flying vertebrates such as birds and bats have high metabolic rates. A high metabolic rate means a high energy output, necessary to sustain flight, but that has to be fuelled by a high intake of energy, in the form of nutritious food, and a high intake of oxygen. To protect the system, some kind of insulation is essential: fur keeps the heat in and saves energy wastage, an especially high risk factor for tiny animals. Insects operate very differently, but large forms, such as moths and bees, have fur for the same reason.

Palaeontologists have debated the origins of pterosaurs since their first discovery. They have always been classified as reptiles of some sort, even though they may resemble bats in certain ways. Like bats, their wings are constructed from modifications to the arms and made from a thin layer of skin, stretched along the arm. However, whereas in bats the wing is in segments, with all fingers extending through the structure, in pterosaurs only the fourth finger is elongate; the other fingers are still short and used

The first attempt to understand what pterosaurs might have looked like, a lively drawing by Edward Newman published in 1843. At the time, Newman thought they were some kind of flying marsupial, but at least the idea that they were covered with insulating fur had already been established.

for climbing and grasping. The wing membrane stretches from the knee and the side of the body to the tip of the fourth finger and, because there aren't other fingers to help hold the wing membrane out, it is stiffened by thin rods of collagen called aktinofibrils that run diagonally from front to back of the wing membrane.

No other animal has a wing like this, so where did pterosaurs fit into the evolutionary tree? The oldest pterosaurs have these wings as well as other specializations, and there is no obvious half-way form between pterosaurs and something else. The answer was established when palaeontologists began to look again at dinosaurs and their ancestors, and to apply cladistic methods, based on identifying shared specialized characters, to the question of reconstructing the evolutionary tree. It turns out that the hind limbs of pterosaurs are very like those of dinosaurs and their immediate ancestors. Some researchers argue that pterosaurs had a much deeper origin in the reptilian evolutionary tree, but most agree that the specialized features of the hindlimb (elongated feet, hinge-like ankle and knee joints, and the ball-and-socket hip joint) are identical between pterosaurs and dinosaurs, and therefore that both groups shared an ancestor.

This means that there is a wider group that includes both dinosaurs and pterosaurs as well as their immediate ancestors, termed collectively the avemetatarsalia, a bit of a mouthful of a name I coined in 1999. The name means 'bird-metatarsals', the metatarsals being the long ankle bones, and all the evidence suggests that the group arose in the Early Triassic. This was a time when ecosystems were in the process of rebuilding after the famous end-Permian mass extinction had wiped out 95% of species, 252 million years ago; the few survivors, on land and in the oceans, inherited a devastated world. But, without the animals and plants that had formed the fixed outlines of pre-extinction ecosystems, the survivors built new kinds of life modes and ecological interactions. Among land animals this included warm-bloodedness of various sorts, faster lifestyles and upright posture.

The ancestral avemetatarsalian, a tiny upright-running quadruped, was fast enough and smart enough to survive and was the ancestor of major Mesozoic dynasties. There is a fossil gap of nearly 20 million years from Early to Late Triassic, when we know there must have been pterosaurs and dinosaurs, but we have not yet found any fossils. This is a challenge I offer to my students – we know they were there, so go out and find these earliest of all dinosaurs and pterosaurs!

Pycnofibres or feathers

So, we have evidence of the pelage of pterosaurs, but what exactly was it? In 1996, David Unwin and Natasha Bakhurina, then both based at the University of Bristol, provided a detailed description of one of the best-preserved examples of pterosaur dermal coverings. This fossil was from the Late Jurassic of Kazakhstan, and evocatively named *Sordes pilosus*, or 'hairy devil'. Unwin and Bakhurina studied the specimen under lighting of various kinds and under the microscope, and they noted various kinds of fibrous structures, including the aktinofibrils of the wings and the long, slender whiskers that insulated the wings and body. These two slender structures are entirely different in composition and function, and it is important not to confuse them.

In 2009, Brazilian pterosaur expert Alexander Kellner named the insulating whiskers of pterosaurs as pycnofibres, meaning 'dense' fibres. Kellner had been studying the exceptionally preserved anurognathid pterosaur *Jeholopterus* from China, and he could see the distinction between whiskers and aktinofibrils very clearly, and realized the whiskers needed a name. Hairs in mammals, feathers in birds, and pycnofibres in pterosaurs were all seen as similar in origin and function – they emerged from follicles in the skin, they were constructed primarily of the protein keratin, and they functioned to create an insulating layer over the body.

Later, palaeontologists identified melanosomes within the pterosaur pycnofibres, apparently providing further evidence that hairs, feathers and pcynofibres are essentially the same structurally. Biologists had also been studying embryos of birds and mammals, and found that the same key regulatory genes turn on the production of feathers in birds and hairs in mammals. This confirmed that hairs and feathers are 'deep homologues', meaning they share an evolutionary origin early in the history of vertebrates. We cannot analyse the genome of pterosaurs, but it seemed likely that pycnofibre growth was regulated similarly in the early embryo.

Then, in a 2019 paper, Baoyu Jiang from the University of Nanjing and my team at the University of Bristol made the rather cheeky assertion that what had been termed pycnofibres are in fact feathers. Our study emerged from two specimens of anurognathid pterosaurs that Baoyu Jiang had obtained during a field excursion to Inner Mongolia in 2017, to which I had been invited. Each slab showed a complete pterosaur skeleton in the rock, the body and head in the centre, and the wings partially folded on each side.

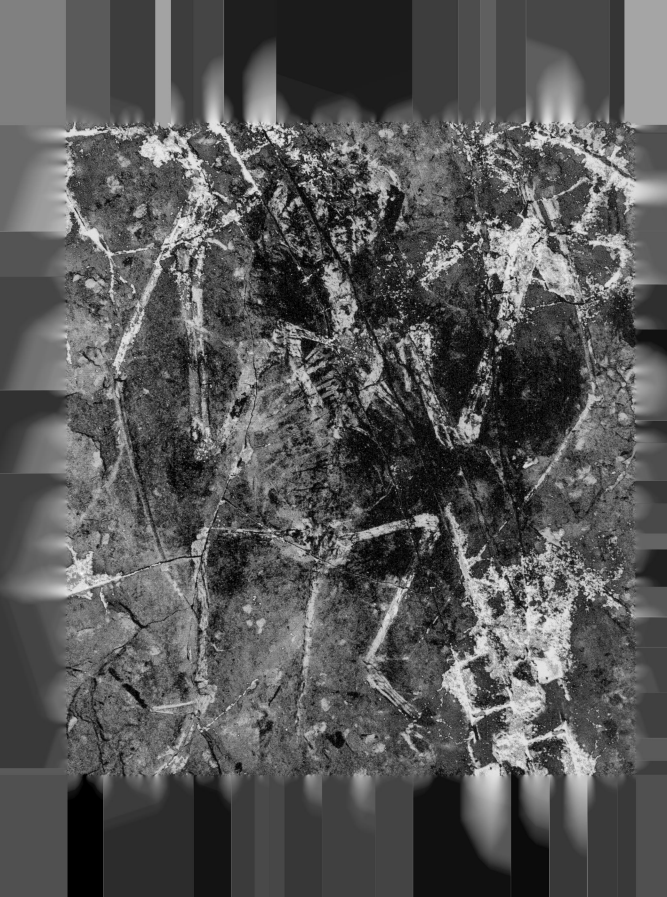

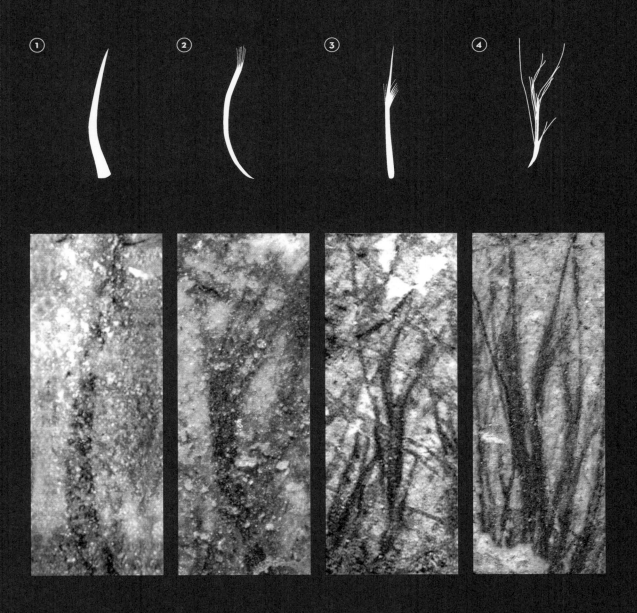

The Chinese anurognathid specimens, showing the complete skeleton with legs and wings folded. Feathers were sampled from all over the head, torso, legs and wings, and they include four feather types, simple filaments (1), tufted monofilaments (2), filaments with tufts halfway along (3), and fluffy down-type feathers (4).

Insect remains littered the slabs. Indeed, I spent a week in the field chipping rocks, and I found numerous insects, small flies, big meaty caddis flies and cockroaches, and delicate lacewings – though not a hint of a pterosaur or a dinosaur.

Jiang's student, Zixiao Yang, studied the pterosaur specimens minutely, revealing pycnofibres all over the head and back and sparsely over the wings. These contained abundant melanosomes, providing evidence that the Yanliao anurognathid was light brown in colour. And they revealed another surprise. Yang distinguished the pycnofibres on the wings from the aktinofibrils, and then the surprise. He identified four kinds of pycnofibres: the simple monofilaments we had expected, and as described by Kellner ten years earlier, as well as three kinds of branching pycnofibres. Some branched at the base, like a bird's down feather, and others had tufts either at the end of the structure, or halfway down. This caused us to think again about pycnofibres, and the fact that many of them showed branching confirmed that these were in fact feathers. (A feather is defined in the dictionary as a branching structure; QED.)

Anurognathids had large eyes and their vision was binocular (three-dimensional). This made them as well adapted to flying as birds are: flyers require three-dimensional vision to make sure they can judge distances well and avoid crashing into trees.

The impact of pterosaur feathers

If dinosaurs and pterosaurs both had feathers, then this would suggest that feathers might have originated with the origin of the Avemetatarsalia. This assertion needs further study, and it would become more convincing if older specimens can be found.

However, the sequence of discoveries of feathered animals since 1996 has shown that feathers, identified at their oldest in *Archaeopteryx*, dated at 150 million years ago, have descended the evolutionary tree. First, feathers were seen to have originated as early as *Sinosauropteryx* (pp. 28–41), some distance below *Archaeopteryx* in evolutionary terms. Then, in 2014, our report of feathers in *Kulindadromeus* (pp. 168–77) moved their point of origin to the origin of dinosaurs. Our 2019 claim of feathers in pterosaurs moves the point of origin down to the root of the avemetatarsalia, some 250 million years ago, 100 million years before *Archaeopteryx*.

Some scholars have suggested that feathers arose several times independently, in theropod dinosaurs, ornithopod dinosaurs and in pterosaurs, but the simpler explanation is a single point of origin, and a subsequent loss of feathers in armoured dinosaurs such as *Borealopelta* (pp. 192–205) and in sauropods such as *Saltasaurus* (pp. 140–53). Either way, feathers evidently evolved first not for flight but for insulation, and possibly, then, for display, and finally for flight.

This is still a hot debate. Palaeontology may be an 'old' science, tracing its origins back 200 years or more, but new fossils can still put the cat among the pigeons.

15

TUPANDACTYLUS

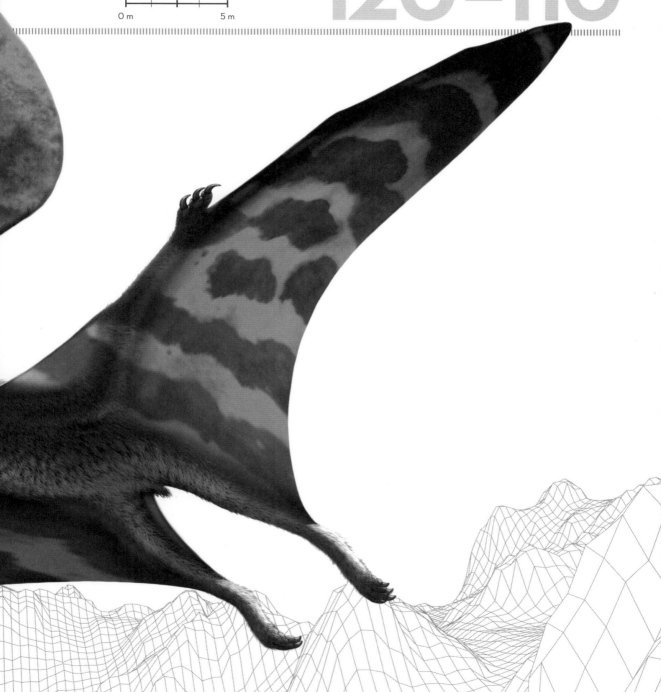

0 m   5 m

Colourful head crests

This is the kind of animal that makes palaeontology such a hit. Who could not look at *Tupandactylus* and say, 'This can't be real; that crazy head shield; it surely couldn't fly. Pure fantasy!' And yet this animal is real, and it is just one of numerous equally unlikely-looking pterosaurs that once soared in our skies.

We are in the Early Cretaceous of Brazil, 113 million years ago. The scene is a tropical lagoon, with warm, shallow seas full of fishes and other marine animals. Large pterosaurs swoop low over the water, occasionally dipping their beaks in the water to grab a fish. The seashore is fringed with ferns, seed ferns and tall conifer trees, especially monkey puzzles. These trees are home to many species of insects, spiders, tree frogs, lizards and birds. Crocodiles swim lazily, snapping at fishes in the shallows.

But it's the pterosaurs that draw the eye. They are huge, with wing spans of up to 5 metres (16 feet), and most of them have improbably large, ungainly heads. *Tupandactylus* is no exception. Its long, toothless jaws and deep throat pouch confirm that it was a fish-eater. One individual skims over the

Reconstruction of the giant pterosaur *Pteranodon* from the Late Cretaceous of North America, taken from the classic work of George Eaton in 1910. Notice how long the wings are in proportion to the size of the torso. However, the long, slender neck and huge, but lightweight, head have always puzzled aerodynamics experts.

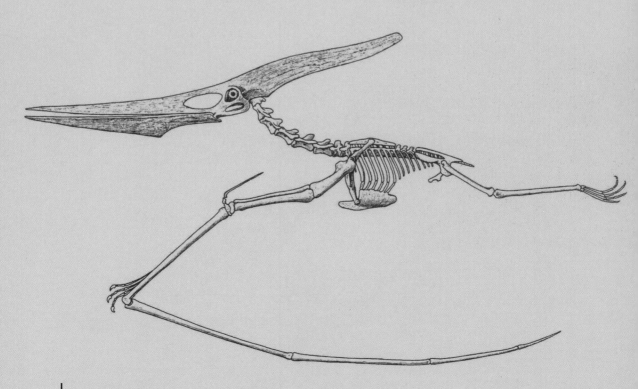

surface of the water, barely twitching its wings, relying on updrafts from the waves to maintain an even height. Its head flicks from side to side as it seeks movements in the water with its beady little eyes. Something stirs in the soft white sand: a flat fish changing its position. *Tupandactylus* inclines its head, trailing the lower jaw in the water as it approaches, then down goes the head, scooping up the sand, and it lifts again with a beat of its wings. Water flows out of the sides of the mouth and the throat pouch fills up. Something is punching and squirming inside the pouch, but with an upward tilt of its head, the squirming moves to the neck, and down it goes. With a slightly plump tummy, the pterosaur flies on, seeking a spot on the shore where it can settle safely and digest its prey.

Tupandactylus's most startling feature is, undoubtedly, the head crest. Whereas the body is covered with short, brown, whiskery feathers, the huge head sail is brightly coloured. It poses a whole series of questions: how can the animal carry this huge thing on its head; why doesn't it somehow prevent it from flying; why is it so brightly coloured?

Giant pterosaurs

Fossils of Cretaceous pterosaurs were found in the south of England from 1840 onwards. Though many were only isolated jaws or wing bones, or fragments of the skull, they hinted already at something much bigger than the Jurassic pterosaurs that had been reported up to that time.

The same story was repeated when the great American palaeontologists Edward Cope and Othniel Marsh began excavating dinosaurs in the midwestern United States (see p. 15). In the 1870s, they both began to identify pterosaur bones, but the fossils were tantalizingly incomplete. It seemed, nevertheless, that they had uncovered something much bigger than the English specimens. A complete skull was found in 1876, and it was named *Pteranodon* by Marsh. This skull was entirely different from those that had been found before; it was a metre (3 feet) in length, with long, deep jaws that ended in a point at the front, but carried no teeth. The eye and brain were tucked into a small space under a long crest that projected backwards from the head. It was the size of this crest that amazed Marsh: how could a pterosaur balance this metre-long hatchet head on its neck, and at that size ever become airborne?

Following pages:
The complete skeleton of *Pteranodon*, wings outstretched, soaring high over the Late Cretaceous Western Interior Seaway.

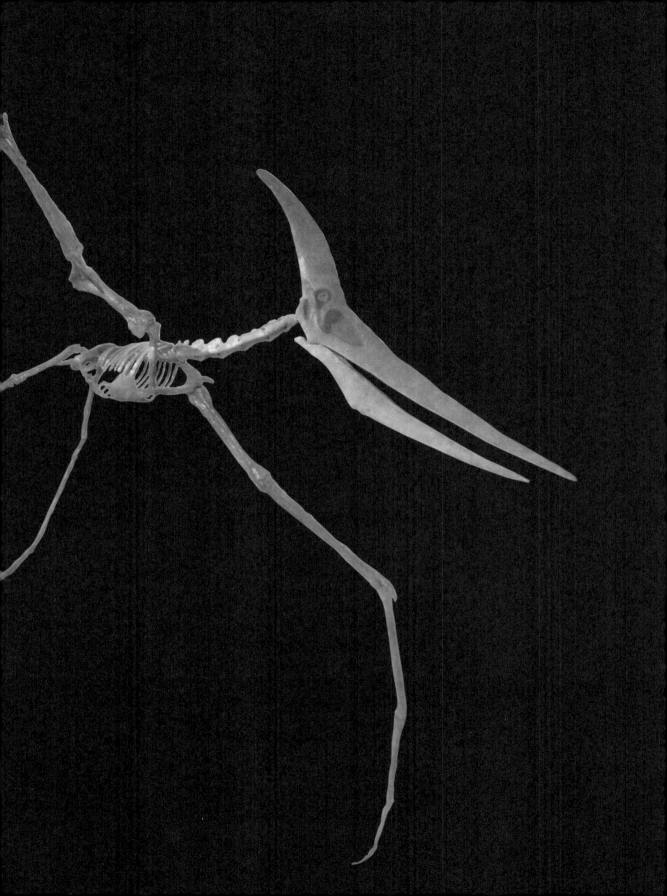

By 1910, enough pieces of the *Pteranodon* skeleton had been found that palaeontologist George Eaton was able to make a complete reconstruction. This and later studies showed that *Pteranodon* females had a wingspan of 3.8 metres (12½ feet), and males 5.6 metres (18½ feet). There were endless debates about the function of the crest – was it to balance the long beak at the front, or did it act like a weather vane, somehow allowing the animal to steer, or to keep its nose straight into the airstream as it flew? Indeed, could *Pteranodon* even fly?

Many biologists looked at *Pteranodon* and declared that it surely could not have flown. After all, the world's largest flying bird today, the albatross, has a wingspan of up to 3.7 metres (12¼ feet), and it struggles to land and take off. Once it is flying, it uses natural air flows to keep aloft without beating its huge wings too much, but landing is awkward, as is taking off. Many people will have seen a goose or swan landing on a lake – it may look graceful enough when it is flying at height, but as it slows down to land, tucking its tail feathers in, feet out straight in front, and then hits the water feet first, it looks like a bad water skier, zig-zagging over the water, using the soles of its feet to slow down, and rocking wildly, and using its wings to prevent itself from keeling over.

And yet *Pteranodon*, often with a wingspan over 5 metres (16½ feet), had wings and a lightweight skeleton; why would it have these adaptations if it did not fly? In a classic paper in 1974, Cherrie Bramwell, an expert on modern bats and bat flight, and G. R. Whitfield, a physicist, from the University of Reading tackled the question head-on. They calculated the body mass and flight characteristics of the pterosaur and concluded that it definitely could fly: 'The mode of life is considered, showing that *Pteranodon* probably lived on sea cliffs facing the prevailing wind. After landing on the top, it would scrabble forwards (it could neither stand up nor walk) and hang from its hind feet over the edge. From here it could easily launch itself. When flying near the cliff it would soar in the hill lift; when far out at sea it would use the weak thermals generated by convection over the warm sea. Dynamic soaring and slope-soaring over the waves are not possible for such a slow-speed glider.' They estimated that a large *Pteranodon* weighed only 16.6 kilograms (36½ pounds), and so could readily support that weight on its huge wings.

In searching for information about Bramwell and Whitfield, I found a charming BBC video on YouTube, dating from 1982, in which a young Cherrie Bramwell talked about her pet fruit bat, named Balls, and how she advertised

Experiments with a *Pteranodon* model in 1974. The biologist Cherry Bramwell and the physicist George Whitfield built life-sized models of *Pteranodon* with a 7-metre (23-foot) wingspan and launched them either by hand, as here, or from a huge catapult. Their aim was first to study the airworthiness of the model, and they tried also to add some power, but the weight of motors and batteries made those experiments impossible.

Opposite:
Mark Witton's famous
reconstruction of a grounded
Quetzalcoatlus. This truly
enormous Late Cretaceous
pterosaur from Texas stood as
tall as a giraffe, and yet must
have weighed only a fraction
of its weight if it were to be
able to fly.

for some female fruit bats to keep him company. She reported that he had since fathered about 100 baby fruit bats. Bramwell was in the habit of wearing Balls around her neck as she went out to parties or to go shopping. Her detailed 1974 paper still stands as a masterpiece of close anatomical study and careful, first-principles analysis of the biomechanics of an extinct animal.

Two years later, in 1976, an even larger pterosaur was reported from Texas. This was *Quetzalcoatlus*, also from the Late Cretaceous, and with an estimated wingspan of 10–11 metres (33–36 feet), twice the size of *Pteranodon*. The initial finds constituted a few isolated elements – a jaw, and some very long neck bones – but more remains found since allow a reconstruction. This pterosaur was as tall as a giraffe when standing on the ground, with its wings folded. Surely *this* amazing creature could not fly?

In 2010, Donald Henderson, an expert in biomechanics of ancient animals and co-discoverer of *Borealopelta* (pp. 192–205), estimated that *Quetzalcoatlus* weighed about 540 kilograms (1,190 pounds), and so clearly could not fly. Others, including Mike Habib and Mark Witton, have estimated the body mass as 200–250 kilograms (440–550 pounds). Their calculations suggest that it could indeed have flown, but was adapted for long-distance, sustained flight, being able to cover thousands of kilometres during continuous flights of seven to ten days. All experts, though, cringe when they are asked to explain how this massive animal could succeed in taking off and landing without wrecking its fuselage.

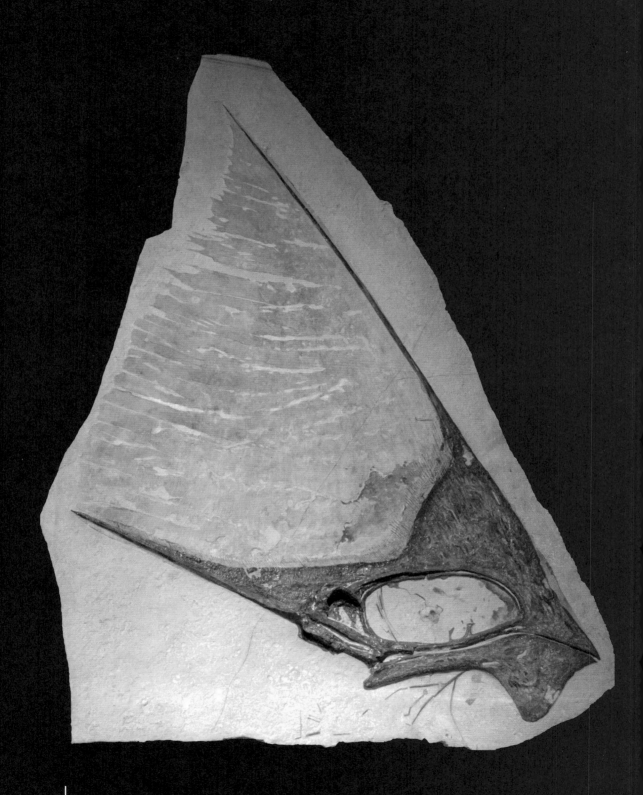

Head crests, melanosomes and sexual flashing

Unlike the earliest pterosaur finds, *Tupandactylus* specimens are often quite complete, because they are found in slowly accumulating limestone sediments, deposited in the Araripe Basin in eastern Brazil. Fossils of marine fishes, as well as the remains of pterosaurs, plants and insects blown into the lagoon from nearby land, fell on the sediment surface and were covered gently by more slowly moving sediment. In some cases, the burial was rapid and soft tissues became preserved, including, for example, the muscles of the fish and pieces of pterosaur skin.

When the crested pterosaurs, such as *Tupandactylus*, were found, they were very hard to interpret. The skulls seemed to consist of a hatchet-shaped lower part, comprising the jaws and eye sockets, and above that a thin, vertically rising bony strut, like the mast of a yacht. Behind this bony strut were sometimes black organic remains, suggesting the former presence of skin. So, the crest was supported by a bony strut at the front, but the rest was simply skin, which would have helped to keep the weight of the structure down.

In a 2011 paper, Brazilian palaeontologist Felipe Pinheiro and colleagues from the University of Rio Grande do Sul described a *Tupandactylus* specimen with traces of pycnofibres found not only around the head, as would be expected (see p. 215), but also on the crest. They noted near-vertical fibres that rose from the snout to the very top of the crest. Indeed, the whole crest sail contained these fibres, which ran more or less vertically and would have stiffened the crest skin, just as aktinofibrils stiffened the wing membrane of the anurognathids (see p. 214). These authors noted two possible crest shapes, one with the posterior margin of the sail concave, and the other concave and bulging out as a rounded margin. These authors also noted definite colour patterns on the crest skin, but the colours were uncertain.

Opposite:
The *Tupandactylus* skull, showing the extraordinary shape. Everything is distorted by the display sail, which was made from skin and supported across the thin splints of bone in front and behind. Even the lower jaw has a remarkable descending flange, and the snout is shortened to maintain a narrow structure that cut through the air as the animal flew, but enabled complex signalling in life.

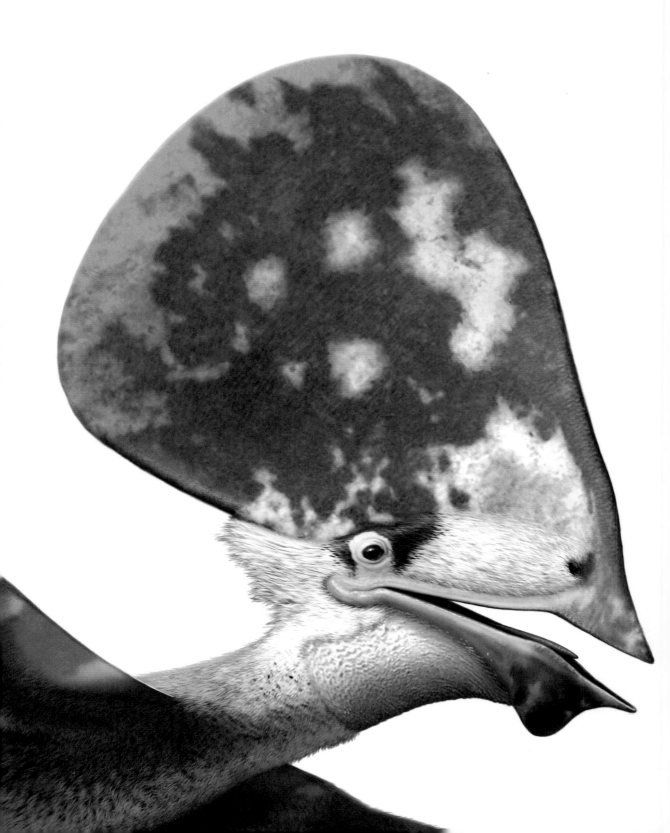

A year later, in 2012, Pinheiro's team reported fossilized bacteria in the same specimen, noting that the soft-tissue crest contained very tiny spherical and rod-shaped granular bodies. They speculated about whether these might be melanosomes, but ultimately concluded that they were not organized in a regular enough way, and so must be decay bacteria. Subsequently, however, various investigators pointed out that these granules resided in the skin, and so were almost certainly melanosomes and indeed provided evidence for the colour of the crest.

Then, in a 2019 paper, Pinheiro's team presented chemical analyses that showed the presence of both eumelanosomes and phaeomelanosomes, but mainly the latter, in the head crest. Their opinion was swayed by the fact the granules were of remarkably even size (bacteria might show large and small examples), there was no evidence of splitting (melanosomes don't split, whereas bacteria split as they grow), and there was no sign of bacterial side-products. The chemical tests showed that the granules retain traces of the melanin pigments, but they are replaced by calcium phosphate, with the phosphate likely mainly remobilized from decaying tissues of the pterosaur body.

In 2020, Cary Woodruff of the University of Toronto, together with Darren Naish and Jamie Dunning, suggested that the *Tupandactylus* crest might even have been photoluminescent. They base this argument on the observation that many modern lizards and birds have structures that can only be seen at ultraviolet (UV) wavelengths, and indeed many animals can see such colours. Some seabirds, such as the puffin and one species of auklet, have photoluminescent structures around their brightly coloured beaks so that other birds, with visual capabilities in the UV spectrum, see the beak structures in a highlighted manner.

In the case of these birds, the photoluminescent structures are expressed in the keratin sheath of the beak or specialized head crests, sometimes called casques. Therefore, Woodruff and colleagues suggested that such prominent head crests and beaks sheathed in keratin as are seen in various dinosaurs and pterosaurs might also have been photoluminescent. In tests on the keratin horn sheaths over the dorsal spines of *Borealopelta* (see pp. 192–205), these authors showed that slender stripes can still photoluminesce in the fossils.

Returning to the Early Cretaceous of north-eastern Brazil, we can imagine the elaborate head crests of *Tupandactylus* flashing different colours at dusk, males and females perhaps showing different patterns, and putting on a spectacular light show in the crepuscular gloom.

Opposite:
The otherworldly glamour of *Tupandactylus*. Flip back a page to see how the very lightweight skull supported the amazing head crest, all of it based on thin layers of skin stretched like a sail from two bony struts. The mottled shades must have signalled important messages to other members of the species, and colours might even have changed seasonally.

FURTHER READING

The following list of books and papers is by no means complete, but offers a way in to the main themes, in addition to the work of many scientists referred to throughout the book. In the case of books, I've chosen recent examples written by scientists for adult readers, so they give a real flavour of the work in the field and the laboratory.

DINOSAURS

Benton, M. J. 2019. *The Dinosaurs Rediscovered: How a scientific revolution is rewriting history* (Thames & Hudson: London and New York). How palaeontologists conduct their scientific studies of dinosaurs.

Benton, M. J. 2015. *Vertebrate Palaeontology*, 4th edition (Wiley: New York). The most widely used student textbook on dinosaurs and all other fossil vertebrates.

Brusatte, S. 2018. *The Rise and Fall of the Dinosaur: A new history of a lost world* (Harper Collins: New York). An exciting account of historical and modern dinosaur discoveries.

Fastovsky, D. E. and D. B. Weishampel. 2021. *Dinosaurs: A concise natural history*, 4th edition (Cambridge University Press: Cambridge). The best standard textbook on dinosaurs.

Naish, D., and P. M. Barrett. 2018. *Dinosaurs: How they lived and evolved* (Natural History Museum: London). Dinosaurs in full colour – history, diversity and palaeobiology.

Norell, M. A. 2019. *The World of Dinosaurs: The ultimate illustrated reference* (University of Chicago Press: Chicago). On collecting and studying dinosaurs, and with insights from the great work at the AMNH.

MESOZOIC BIRDS, PTEROSAURS AND MAMMALS

Kemp, T. S. 2005. *The Origin and Evolution of Mammals* (Oxford University Press: Oxford). Classic review of mammal evolution.

Kielan-Jaworowska, Z., R. L. Cifelli and Z.-X. Luo. 2004. *Mammals from the Age of Dinosaurs: Origins, evolution, and structure* (Columbia University Press: New York). A few years out of date, but still the most comprehensive overview of Mesozoic mammals.

Pickrell, J. 2014. *Flying Dinosaurs: How fearsome reptiles became birds* (UNSW Press: Sydney). The scientific stories behind the discoveries.

Witton, M. P. 2013. *Pterosaurs: Natural history, evolution, anatomy* (Princeton University Press: Princeton). The most comprehensive book on pterosaurs, with great illustrations.

Xu, X., Z. Zhou, R. Dudley *et al.* 2014. 'An integrative approach to understanding bird origins', *Science* 346: 1253293. A current review of bird origins and the 'trees down' model for the origin of flight.

HISTORY OF DINOSAUR STUDIES AND PALAEOART

Cadbury, D. 2010. *The Dinosaur Hunters: A true story of scientific rivalry and the discovery of the prehistoric world* (Fourth Estate: London). The early days of dinosaur hunting in England.

Dingus, L. *The Dinosaur Hunters: The extraordinary story of the discovery of prehistoric life* (American Museum of Natural History: New York). Episodes in the history supported by reproduction documents.

Rieppel, L. 2019. *Assembling the Dinosaur: Fossil hunters, tycoons, and the making of a science* (Harvard University Press: Cambridge, Mass.). How the early discoveries were made, and all the characters involved.

White, S. 2012, 2017. *Dinosaur Art: The world's greatest paleoart*, 2 volumes (Titan Books: New York). Highly illustrated selection of dinosaur representations over the decades.

Witton, M. 2018. *The Palaeoartist's Handbook: Recreating prehistoric animals in art* (Crowood Press: Marborough). How to reconstruct dinosaurs, by a successful palaeoartist.

DYNAMIC DINOSAURS

Alexander, R. McN. 1989. *Dynamics of Dinosaurs and Other Extinct Giants* (Columbia University Press: New York). Still an excellent introduction to palaeobiology.

Alexander, R. McN. 1976. 'Estimates of speeds of dinosaurs', *Nature* 261: 129–30. The first numerical and testable study of dinosaur speeds.

Bakker, R. T. 1986. *The Dinosaur Heresies: New theories unlocking the mystery of the dinosaurs and their extinction* (W. Morrow: New York; Longman: Harlow). The first, controversial proposals that dinosaurs were warm-blooded, lively animals.

Barrett, P. M. and E. J. Rayfield. 2006. 'Ecological and evolutionary implications of dinosaur feeding behaviour', *Trends in Ecology and Evolution* 21: 217–24. A review of how palaeontologists determine dinosaur feeding behaviour.

Erickson, G. M. 2005. 'Assessing dinosaur growth patterns: a microscopic revolution', *Trends in Ecology and Evolution* 20: 677–84. Fast growth rates in dinosaurs.

Hutchinson, J. R. and S. M. Gatesy. 2006. 'Dinosaur Locomotion: Beyond the bones', *Nature* 440: 292–94. Using computational visualization to estimate dinosaur running postures and speeds.

Ostrom, J. H. 1969. 'Osteology of *Deinonychus antirrhopus*, an unusual theropod from the Lower Cretaceous of Montana', *Bulletin, Peabody Museum of Natural History* 30: 1–165. The classic description of *Deinonychus*, and the paper that showed birds evolved from dinosaurs.

Rayfield, E. J. 2007. 'Finite element analysis and understanding the biomechanics and evolution of living and fossil organisms', *Annual Review of Earth and Planetary Sciences* 35: 541–76. A review of the FEA method as applied to dinosaurs and other fossil animals.

Sander, P. M. *et al.* 2010. 'Biology of the sauropod dinosaurs: the evolution of gigantism', *Biological Reviews* 86: 117–55. How to be truly huge and successful.

MODES OF BEHAVIOUR OF FEATHERED DINOSAURS

Palmer, C. 2014. 'The aerodynamics of gliding flight and its application to the arboreal flight of the Chinese feathered dinosaur *Microraptor*', *Biological Journal of the Linnean Society* 113: 828–35. *Microraptor* could probably fly, tested with models in wind tunnels.

Pei, R. *et al.* 2020. 'Potential for powered flight neared by most close avialan relatives, but few crossed its thresholds', *Current Biology* 30: 4033–46. Numerical confirmation that *Microraptor,* like birds, could engage in powered flight.

THE AMAZING FOSSILS FROM CHINA

Chen, P., Z. Dong and S. Zhen. 1998. 'An exceptionally well-preserved theropod dinosaur from the Yixian Formation of China', *Nature* 391: 147–52. The first description of a feathered dinosaur, in English, presenting *Sinosauropteryx* to the world.

Chiappe, L. M. and Q. J. Meng. 2016. *Birds of Stone: Chinese avian fossils from the age of dinosaurs* (Johns Hopkins University Press: Pittsburgh). All those amazing fossil birds with feathers preserved.

Ji, Q. *et al.* 2002. 'The earliest known eutherian mammal', *Nature* 416: 816–22. The first specimen of *Eomaia*, then the oldest modern-type mammal.

Ji, Q. *et al.* 1998. 'Two feathered dinosaurs from northeastern China', *Nature* 393: 753–61. *Caudipteryx*, the first dinosaur to be described with pennate feathers.

Long, J. A. and P. Schouten. 2009. *Feathered Dinosaurs: The origin of birds* (Oxford University Press: Oxford and New York). Spectacular illustrations and the importance of the new fossils from China.

Meng, J. 2014. 'Mesozoic mammals of China: implications for phylogeny and early evolution of mammals', *National Science Review* 1: 521–42. Review of finds of early mammals.

Xing, L. *et al.* 2016. 'A feathered dinosaur tail with primitive plumage trapped in mid-Cretaceous amber', *Current Biology* 26: 3352–60. A dinosaur in amber – who would have thought it possible?

Xu, X., Z. Zhou and X. Wang. 2000. 'The smallest known non-avian theropod dinosaur', *Nature* 408: 705–8. The naming of *Microraptor*, the flying four-winged dinosaur.

Yang, Z.-X. *et al.* 2019. 'Pterosaur integumentary structures with complex feather-like branching', *Nature Ecology & Evolution* 3: 24–30. First report of branching feathers in pterosaurs.

MELANIN AND COLOUR IN MODERN ANIMALS

D'Alba, L. and M. D. Shawkey. 2019. 'Melanosomes: biogenesis, properties, and evolution of an ancient organelle', *Physiological Reviews* 99: 1–19. The latest on the colour-generating organelle in feathers and hairs.

Diamond, J. and A. B. Bond. 2013. *Concealing coloration in animals* (Harvard University Press: Cambridge, Mass.). Camouflage functions in modern animals.

Hill, G. E. and K. J. McGraw. 2006. *Bird coloration*, 2 volumes (Harvard University Press: Cambridge, Mass.) The most comprehensive account.

Stevens, M. and S. Merilaita. 2011. *Animal Camouflage: Mechanisms and function* (Cambridge University Press: Cambridge). Collection of papers on camouflage.

COLOUR IN DINOSAURS

Godefroit, P. *et al.* 2014. 'A Jurassic ornithischian dinosaur from Siberia with both feathers and scales', *Science* 345: 451–55. First non-theropod dinosaur with all-over feathers.

Li, Q. *et al.* 2018. 'Elaborate plumage patterning in a Cretaceous bird', *PeerJ* 6: e5831. The appliance of science to analysis of feather colours and chemistry in *Confuciusornis*.

Li, Q. *et al.* 2012. 'Reconstruction of *Microraptor* and the evolution of iridescent plumage', *Science* 335: 1215–19. The first report of fossilized iridescent plumage.

Li, Q. *et al.* 2010. 'Plumage color patterns of an extinct dinosaur', *Science* 327: 1369–72. The Yale group show the colours and patterns of feathers in the Jurassic dinosaur *Anchiornis*.

Roy, A. *et al.* 2020. 'Recent advances in amniote palaeocolour reconstruction and a framework for future research', *Biological Reviews* 95: 22–50. Telling the colour of dinosaur feathers and other fossils.

Vinther, J. 2020. 'Reconstructing vertebrate paleocolor', *Annual Reviews of Earth and Planetary Sciences* 48: 345–75. Review of the state of the art.

Vinther, J. *et al.* 2016. '3D camouflage in an ornithischian dinosaur', *Current Biology* 26: 2456–62. First proof of countershading in a dinosaur.

Vinther, J. *et al.* 2008. 'The colour of fossil feathers', *Biology Letters* 4: 522–25. The case for melanosomes in fossil feathers.

Zhang, F. *et al.* 2010. 'Fossilized melanosomes and the colour of Cretaceous dinosaurs and birds', *Nature* 463: 1075–78. Our paper in which we show *Sinosauropteryx* had a ginger and white stripy tail.

ORGANIC MOLECULES AND EXCEPTIONAL PRESERVATION

Barbi, M. *et al.* 2019. 'Integumentary structure and composition in an exceptionally well-preserved hadrosaur (Dinosauria: Ornithischia)', *PeerJ* 7: e7875. Detailed study of a hadrosaur mummy's skin and preservation.

Brown, C. M. *et al.* 2017. 'An exceptionally preserved three dimensional armored dinosaur reveals insights into coloration and cretaceous predator–prey dynamics', *Current Biology* 27: 2514–21. Description of the ankylosaur *Borealopelta* and its skin and armour.

Lindgren, J. *et al.* 2014. 'Skin pigmentation provides evidence of convergent melanism in extinct marine reptiles', *Nature* 506: 484–88. Evidence that ichthyosaurs had black-coloured stomachs.

McNamara, M. E. *et al.* 2018. 'Fossilized skin reveals coevolution with feathers and metabolism in feathered dinosaurs and early birds', *Nature Communications* 9: 2072. The famous find of dinosaur dandruff: evidence they didn't shed their skin whole, like lizards and snakes.

Manning, P. L. 2009. *Grave Secrets of Dinosaurs: Soft tissues and hard science* (National Geographic Society: Washington). Everything about dinosaur mummies and their scientific study.

Manning, P. L. *et al.* 2009. 'Mineralized soft-tissue structure and chemistry in a mummified hadrosaur from the Hell Creek Formation, North Dakota (USA)', *Proceedings of the Royal Society B* 276: 3429–37. Microscopic study and chemistry of the North Dakota hadrosaur mummy.

Pinheiro, F. L. *et al.* 2019. 'Chemical characterization of pterosaur melanin challenges color inferences in extinct animals', *Scientific Reports* 9: 15947. The supposed melanosomes in the *Tupandactylus* head crest.

INDEX

Page numbers in *italics* refer to illustrations.

ILLUSTRATION CREDITS

a = above, b = below

8–9 Library of Congress, Washington, D.C.; 10–11 Photo Paolo Verzone/Agence VU; 12–13 Wellcome Collection, London; 14 Smithsonian Institution Archives, Washington, D.C.; 16–17 American Museum of Natural History, New York; 19 Courtesy Peter Galton; 21 Emily Rayfield/Science Photo Library; 31 Raju Soni/Shutterstock; 32–33 Courtesy Fiann Smithwick; 37 © Bob Nicholls; 39, 40 Photo Louis O. Mazzatenta/National Geographic Images; 44 schankz/Shutterstock; 45 Tom Meaker/EyeEm/Getty Images; 46 Martin Shields/Science Photo Library; 47 Photo Diego Delso; 49 Jean-Denis Joubert/Getty Images; 51 Photo Leicester University, LEIUG 115562; 55 Jakob Vinther; 59 © Scott Hartman; 61a Image no. 410765, AMNH Library, New York; 61b Image no. 314661, AMNH Library, New York; 62 Frans Lanting, Mint Images/Science Photo Library; 64–65 Martin Shields/Science Photo Library; 67, 68 Photo American Museum of Natural History, New York; 72 Martin Shields/Science Photo Library; 73 © Scott Hartman; 74–75 Dr. Jingmai Kathleen O'Connor; 77 Matthew Shawkey; 79 Photo Avalon/Universal Images Group via Getty Images; 80–81 Melvyn Yeo/Science Photo Library; 87 Quagga Media/Alamy Stock Photo; 89 Cambridge University Library; 91 Natural History Museum, London/Diomedia Images; 92 Photo Louis O. Mazzatenta/National Geographic Images; 93 From L. Brent Vaughan, *Hill's Practical Reference Library Volume II, 1906*; 96 Roger Harris/Science Photo Library; 100, 102 John Sibbick/Science Photo Library; 103 Photo Louis O. Mazzatenta/National Geographic Images; 104 Quanguo Li, University of Geosciences; Beijing, Keqin Gao, Peking University; Julia A. Clarke Texas University; Matthew D. Shawkey Ghent University; 107 Photo Stephanie Abramowicz/Los Angeles County Museum of Natural History; 108 Photo Louis O. Mazzatenta/National Geographic Images; 110 Quanguo Li, University of Geosciences; Beijing, Keqin Gao, Peking University; Julia A. Clarke Texas University; Matthew D. Shawkey Ghent University; 113 Raimund Kutter/imageBROKER/Shutterstock; 116–17 James Kuether/Science Photo Library; 119 Millard H. Sharp/Science Photo Library; 120, 121 Natural History Museum, London/Diomedia Images; 123 Image no. 330491, AMNH Library, New York; 129, 130 Courtesy of Zhexi Luo, University of Chicago; 132–33 Nature Picture Library/Alamy; 135 Mary Evans/Natural History Museum, London/Diomedia Images; 139 Mike James/Science Photo Library; 142 Dorling Kindersley/UIG/Science Photo Library; 143 Chris R Sharp/Science Photo Library; 144, 147 Ignacio A. Cerda; 148–49 Rebecca Jackrel/Alamy; 151a Steve Gschmeissner/Science Photo Library; 151b Ignacio A. Cerda; 157 De Agostini Picture Library/Diomedia Images; 158 Photo Louis O. Mazzatenta/National Geographic Images; 161 Pascal Goetgheluck/Science Photo Library; 163 Chris Rogers; 164–65 Millard H. Sharp/Science Photo Library; 166 John Serrao/Science Photo Library; 172–73 Photo Royal Belgian Institute of Natural Sciences, Raphus SPRL; 176a, 176b, 177a, 177b Photo Royal Belgian Institute of Natural Sciences; 181 Pascal Goetgheluck/Science Photo Library; 183 Paläontologisches Museum, Munich; 184 Dirk Wiersma/Science Photo Library; 186–87 Natural History Museum, London/Science Photo Library; 189 Museum of Natural History, Oxford; 190–91 Phil Degginger/Carnegie Museum/Science Photo Library; 196–97 Veronique de Viguerie/Getty Images; 199a, 199b Photo Royal Tyrrell Museum, Alberta; 203 Natural History Museum, London/Diomedia Images; 204–5 Photo Royal Tyrrell Museum, Alberta; 209 Sinclair Stammers/Science Photo Library; 210 B.G. Thomson/Science Photo Library; 213 Natural History Museum, London/Alamy; 216–17 Professor Baoyu Jiang and Mr. Zixiao Yang, Nanjing University; 222 From George Eaton, *The Osteology of Pteranodon, 1910*; 224–25 Herve Conge, ISM/Science Photo Library; 227 Photo Brian Wickins; 229 © Mark Witton; 230 American Museum of Natural History, New York

First published in the United Kingdom in 2021 by
Thames & Hudson Ltd, 181A High Holborn, London WC1V 7QX

First published in the United States of America in 2021 by
Thames & Hudson Inc., 500 Fifth Avenue, New York, New York 10110

Designed by Wayne Blades

British Library Cataloguing-in-Publication Data
A catalogue record for this book is available from the British Library

Library of Congress Control Number 2021934195

ISBN 978-0-500-05219-8

Printed in China by RR Donnelley

Be the first to know about our new releases,
exclusive content and author events by visiting
thamesandhudson.com
thamesandhudsonusa.com
thamesandhudson.com.au